How it Works...
How to Fix it
PLUMBING SYSTEMS

Rand McNally & Company

Contents

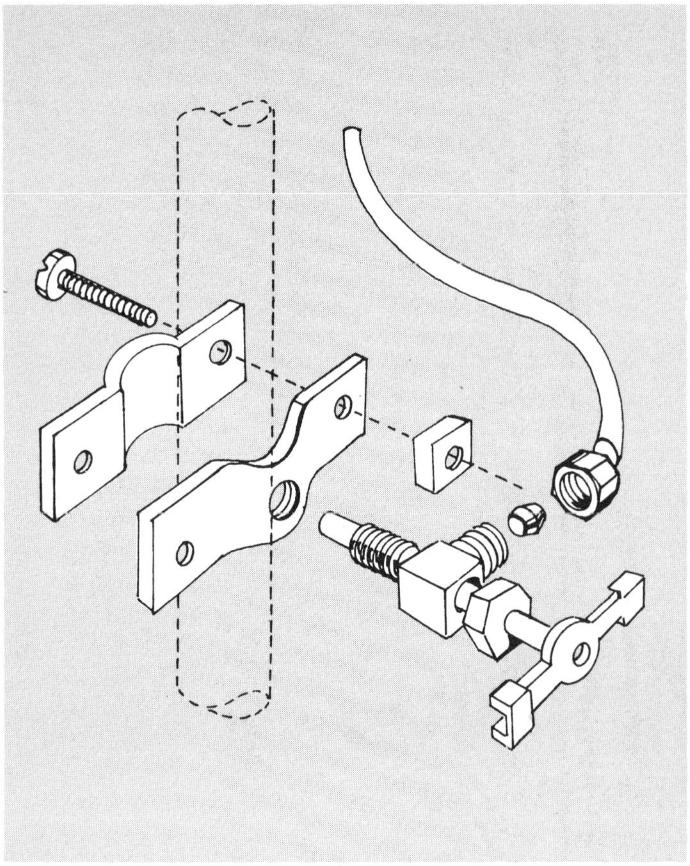

Chapter One

How It Works

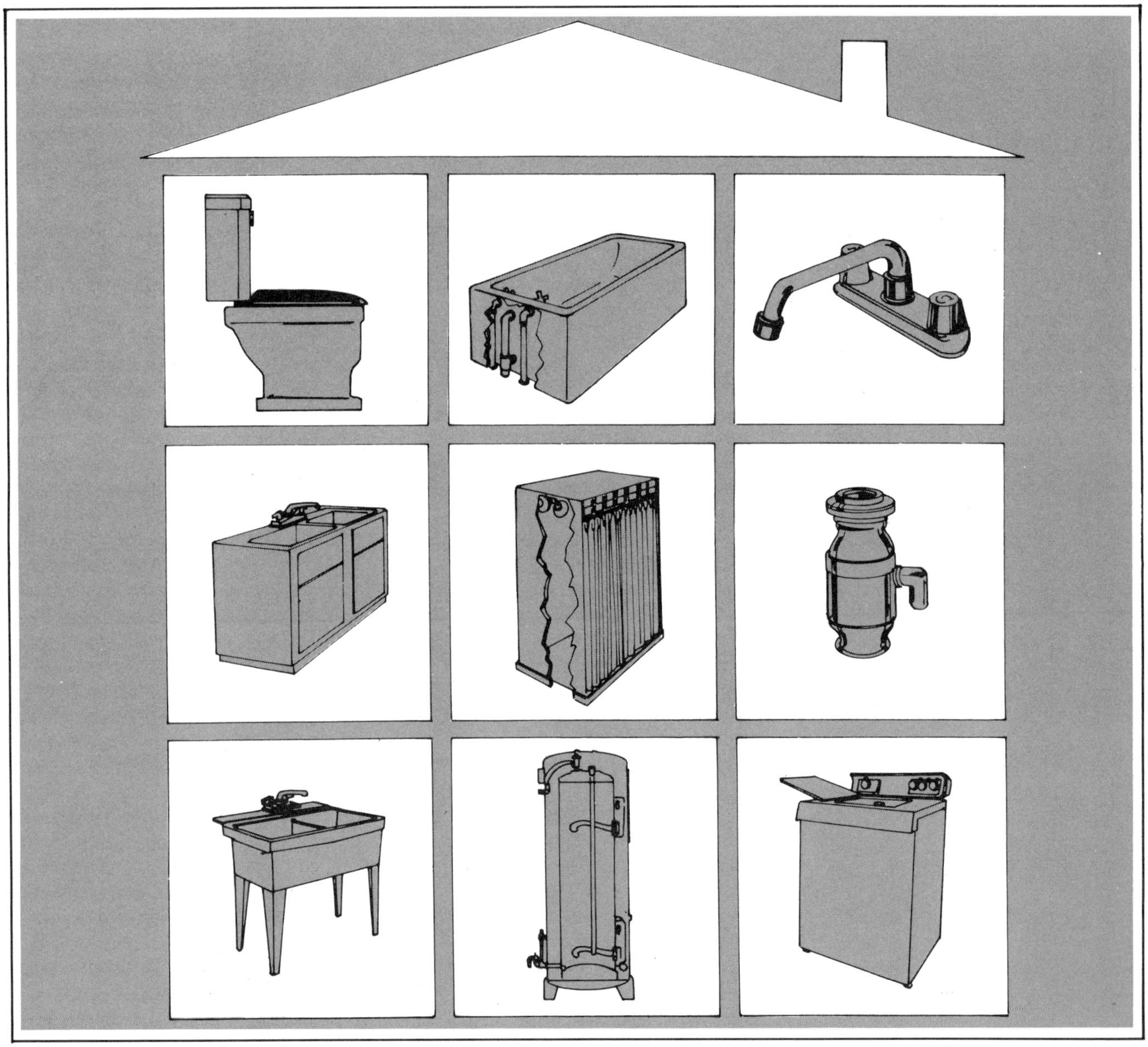

"Plumbing" — few words can strike such terror in the hearts of homeowners and apartment-dwellers. Cartoonists have played upon these fears for years, picturing the hapless victim of a plumbing calamity floating along on the rising waters of his flooding basement. Naturally, there is some truth in all this. Severe plumbing emergencies do arise, and there is probably no individual so lucky as never to be plagued by some type of plumbing problem.

Most likely, however, it is not really the dire events that trouble most people. Their feelings are rooted, instead, in the fear of the unknown. Plumbing to them is a mysterious system, a system which works in ways known only to those who belong to the union and who never seem to be available when you need them the most. In actuality, plumbing works according to some basic laws of nature — gravity, pressure, water seeking its own level — that the rankest amateur can understand in no time. There is little mystery regarding how your plumbing system works, but there is a great deal of misunderstanding. Once you master the basics, you can take care of a great many plumbing problems yourself — and save yourself time, trouble, and money in the process.

SUPPLY AND DISPOSAL

The plumbing in your home is one system composed of two complementary but entirely separate subsystems. One subsystem brings fresh water in, and the other subsystem takes waste water out. Naturally, if you and your family are to stay healthy, there must never be any cross-connection between the supply and disposal subsystems.

If you live in an urban or suburban area, your fresh water is probably pumped to you by the community water department, and your waste water is carried off via a main sewer line to a collective sewage-treatment facility. If you live in a fairly remote or rural area, however, there is a good chance that you depend upon a private well for your water supply and a septic tank for waste disposal. As you might expect, the collective systems for supply and drainage are more efficient than the private systems. That is not to say that the private facilities do not work well; rather, it means that more people can be served at less cost when they are linked to a central community water department and sewage-treatment plant.

SUPPLY SYSTEM

Water that comes into your home does so under pressure. It makes no difference whether your water supply is public or private; water that comes from either a storage tank or a well enters your home under enough pressure to allow it to travel upstairs, around corners, or wherever else it is needed.

In urban and suburban areas, the community water department pumps water into a tank or tower that is higher than the surrounding homes. Then, the force of gravity supplies the water pressure. If you have your own well, then you probably have a pump to bring water into your home under pressure. Occasionally, homes linked to a central municipal supply suffer from insufficient water pressure, and these homes must then add a pump just like those used for private wells.

Water leaves the tank or tower and travels through a water main to the individual supply line for your home. Before you actually can use the water, however, for drinking, bathing, washing clothes and dishes, etc., it must pass through a meter which registers the number of gallons you consume. The meter may be underground somewhere on your property outside your house, or it may be inside your house at the point where the supply pipe enters. The outside type of meter has a metal cover over it; you can lift the cover for access to the meter.

Even if you never care to look at the meter itself, you should be very familiar with one device that is generally located close to the meter. That is

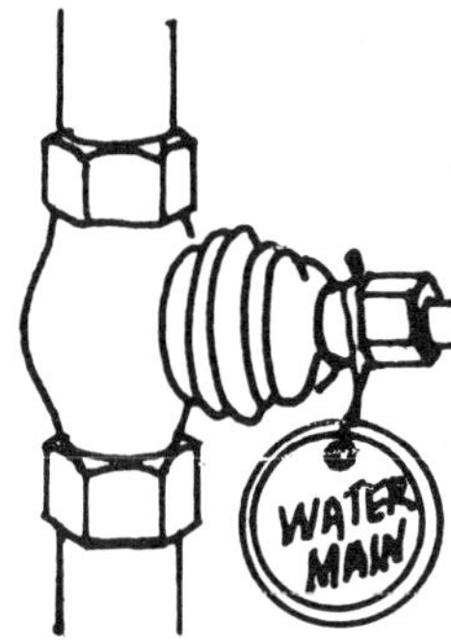

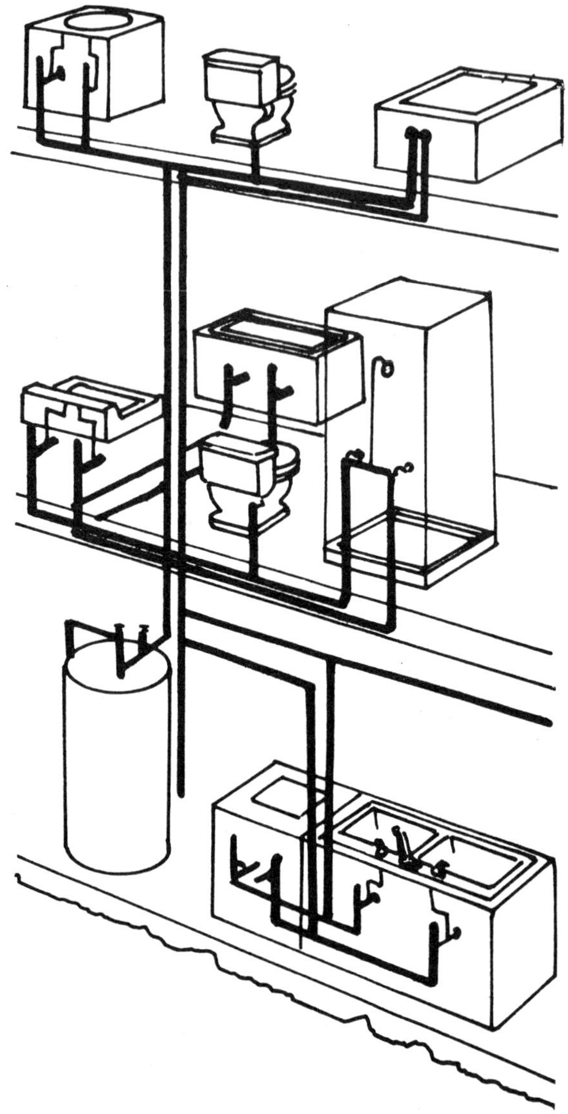

All the water that comes into your home enters under pressure. The pressure permits the fresh water to travel to all the fixtures (right) in your plumbing system. In an emergency, however, you must shut off the supply either at the main cutoff valve (left, above) or at an individual fixture cutoff valve (left, below) or else the water will flood your home.

the main cutoff valve. In case of a plumbing emergency, it is vital that you be able to get to and close off the main cutoff valve as fast as you possibly can. When a pipe bursts, for example, it can flood your house in no time; but you can minimize damage by turning the main cutoff valve which shuts off all the water coming into the house. The cutoff may be a stop and waste valve that drains the water from the pipes as well as shuts off the supply. Naturally, if the emergency is confined to a sink, tub, or toilet, you may not want to stop the entire water supply. Therefore, most fixtures have — or should have — individual cutoff valves.

Once the water has passed from the public facility's tower or tank through the main supply line to your individual supply line and through your home's water meter and main cutoff valve, it is ready to travel to the different fixtures in the house for your use. It is ready, that is, for all your cold water needs. All the water that enters your home is, of course, cold water. As such, it is piped directly to all fixtures that use unheated water through the cold water main; offshoots from the main to the individual fixtures are called branches. The pipes that run vertically, extending upward a story or more, are called risers.

One pipe carries water from the cold water system to your water heater. Then, from the heater a hot water main carries the heated water to all the fixtures, outlets, and appliances that require hot water. You can adjust the degree of heat by raising or lowering the temperature setting on the heater. A thermostat maintains the temperature you select by

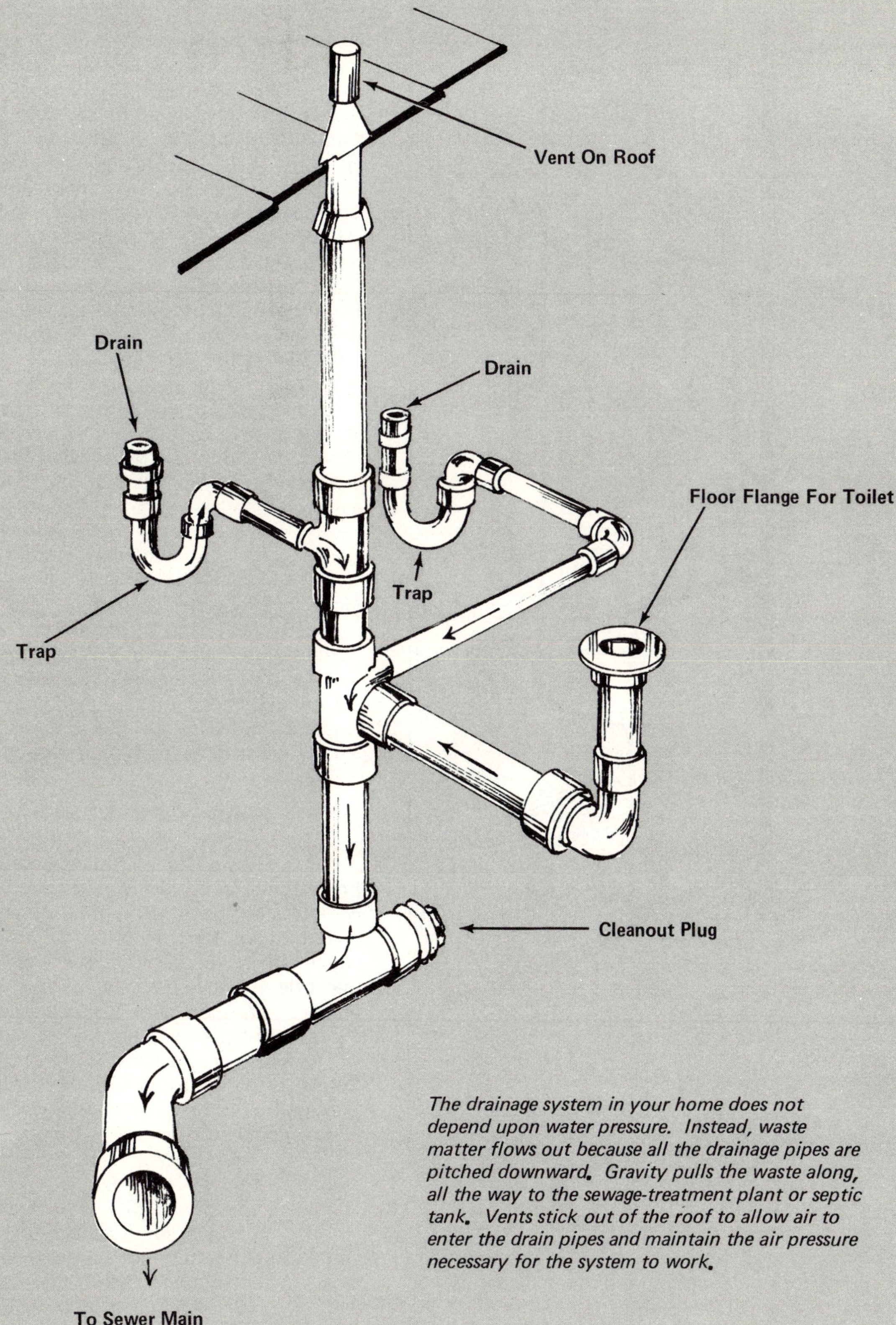

The drainage system in your home does not depend upon water pressure. Instead, waste matter flows out because all the drainage pipes are pitched downward. Gravity pulls the waste along, all the way to the sewage-treatment plant or septic tank. Vents stick out of the roof to allow air to enter the drain pipes and maintain the air pressure necessary for the system to work.

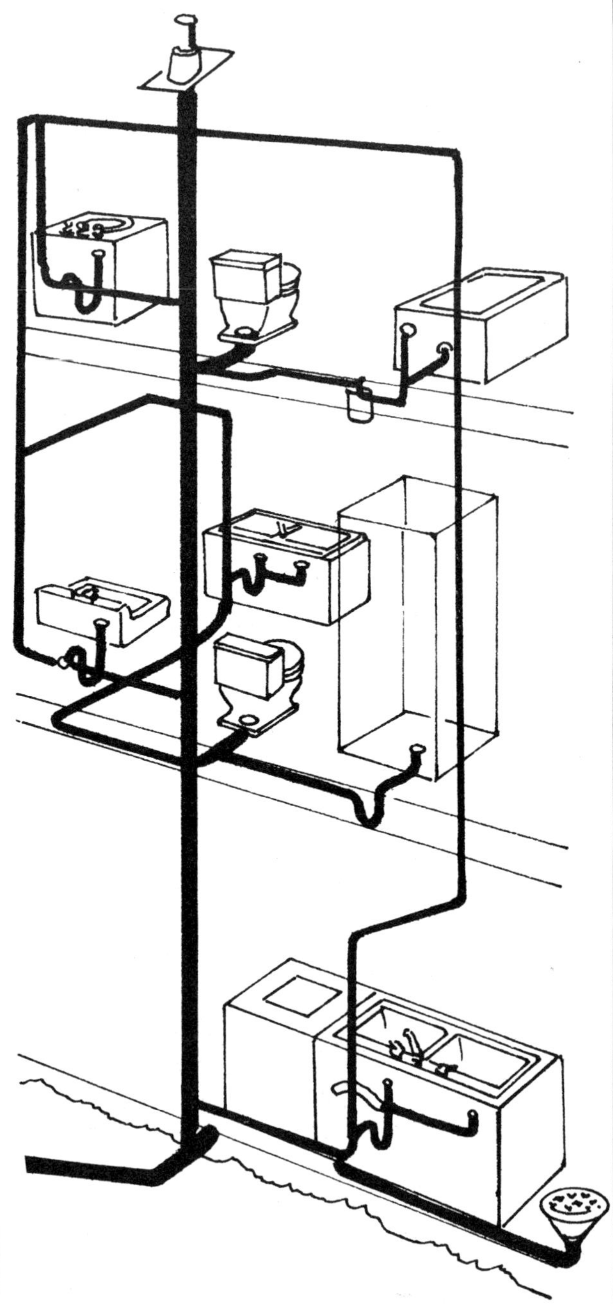

The drainage subsystem of your home's overall plumbing system links every tub, sink, toilet, etc., to a central waste disposal facility. Branch pipes carry the waste matter to the main sewer line.

turning the heating elements on and off as required. The normal temperature setting for a home water heater is 140 to 150 degrees Fahrenheit.

The water pressure in your home is like any other good thing: too much of it can be very bad indeed. Residential water pressure that reaches or exceeds 60 psi (pounds per square inch) can cause your pipes to bang, and it can even cause pipe joints and connections to break. At the very least, excessive pressure wastes water.

You can measure the water pressure in your house by attaching a pressure gauge to any threaded faucet. Be sure to test the pressure at several different times during the day to find a good average. Water pressure does fluctuate, but it should not fluctuate greatly or you will have an uncomfortable time trying to take a shower. One other thing to remember when measuring the water pressure: make sure that no water is running from any other outlet in your home besides the one to which the gauge is attached.

If the gauge registers 60 psi or more, you should install a pressure-reducing valve. A pressure-reducing valve is a fairly inexpensive device, and any do-it-yourselfer can install one easily. Most of these valves work best when installed on a horizontal pipe; the valves come with union fittings that connect to the pipes without difficulty. Once you get it installed, merely set the valve to provide the water pressure that best suits your needs. The valve will lower the pressure and then maintain it at that setting.

Decreasing excessive water pressure is easy; increasing inadequate water pressure is, however, much more difficult. You can tame the 60½ psi flow for the relatively small cost of a pressure-reducing valve, but trying to increase the pressure could require such major endeavors as building your own water tower, installing a pump, or even ripping out all the pipes in your home and installing new ones. Obviously, too little water pressure is something most people learn to accept and live with.

DRAINAGE SYSTEM

Just as the fresh water supply can come from either public or private facilities, the waste water drainage can go to either a public sewer line or private septic tank. Like the public supply facilities, collective sewer systems are much more efficient than private waste disposal methods. After the waste drains from individual houses into a network of pipes, it is carried to a sewage-treatment plant where it is aerated to hasten bacteriological breakdown. Solids are settled out and used for fertilizer, while the liquid is chlorinated and discharged into natural water supplies like streams, rivers, and lakes.

A septic tank and disposal field are designed to handle the waste of a single home. The tank, like the public facility, separates solids from liquid. The solids settle to the bottom of the tank, while the liquid runs out into the disposal field. The tank, of course, must be cleaned every few years, and the disposal field — composed of underground trenches or pits — must be enlarged as it becomes clogged with waste matter that did not settle out completely in the tank.

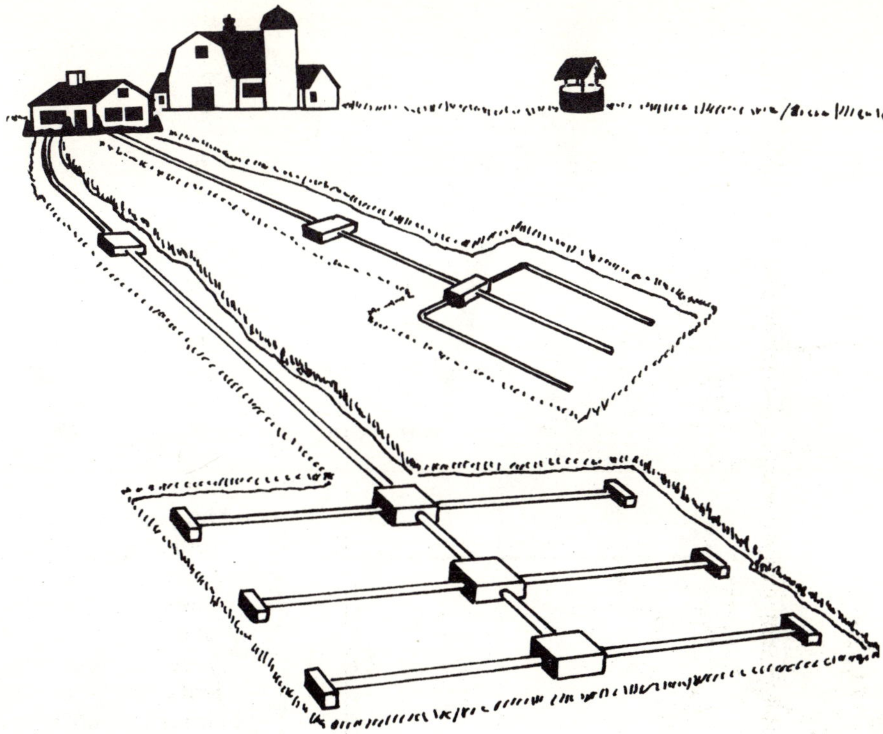

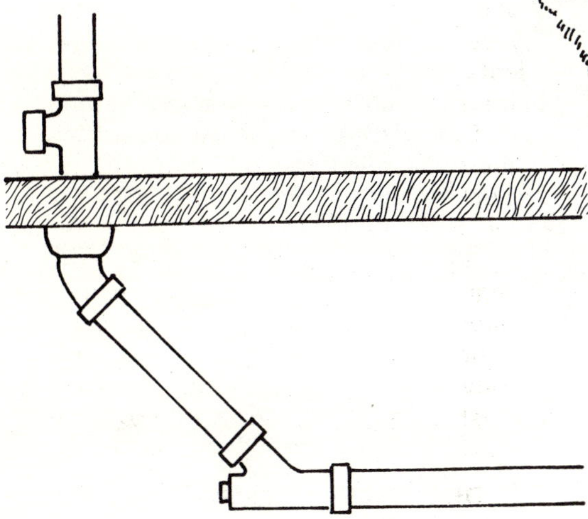

Disposal field (above) of a septic tank system is designed to allow liquid waste to be absorbed into the earth in a controlled manner. Solid wastes settle in the tank and must be cleaned out every few years. Since the main drain pipe (left) is often where clogs occur, it is equipped with cleanout plugs that give you a direct shot at the blockage.

No matter which method of ultimate waste disposal is employed, the drainage systems in homes are essentially the same. Drainage systems do not depend on pressure as do supply systems. Instead, the waste matter leaves your home because the drainage pipes all point downward and gravity pulls the waste along. The sewer line continues the downward flow to the sewage-treatment facility or to the septic tank.

While the system sounds simple — and it is — there is more to waste drainage than pipes tilted downward. There are also vents, traps, and clean-outs. You can see the vents sticking out of the roof of your house. They are there to allow air to enter the drain pipes. If there were no air pressure coming from the vents, then the waste water would not flow out properly and the water in the traps could get siphoned away.

Traps, of course, are vital components of the drainage system. You can see a trap under the drain of every sink in your house; it is the curved or S-shaped section of pipe. Water flows from the basin with enough force to go through the trap and on out through the drain pipe, but enough water stays in the trap to form a seal against the backing up of sewer gas into your home. If there were no seal there, bad odors and dangerous gases could enter your home.

Every fixture must have a trap. Toilets are self-trapped; they do not require an additional trap at the drain. Bathtubs frequently have drum traps which not only form a seal against sewer gas but also collect hair and dirt to prevent clogged drains. Grease traps in some kitchens are similar in that they collect grease that goes down the sink and which might otherwise cause clogging. Because grease and hair are generally the causes of drain clogs, these

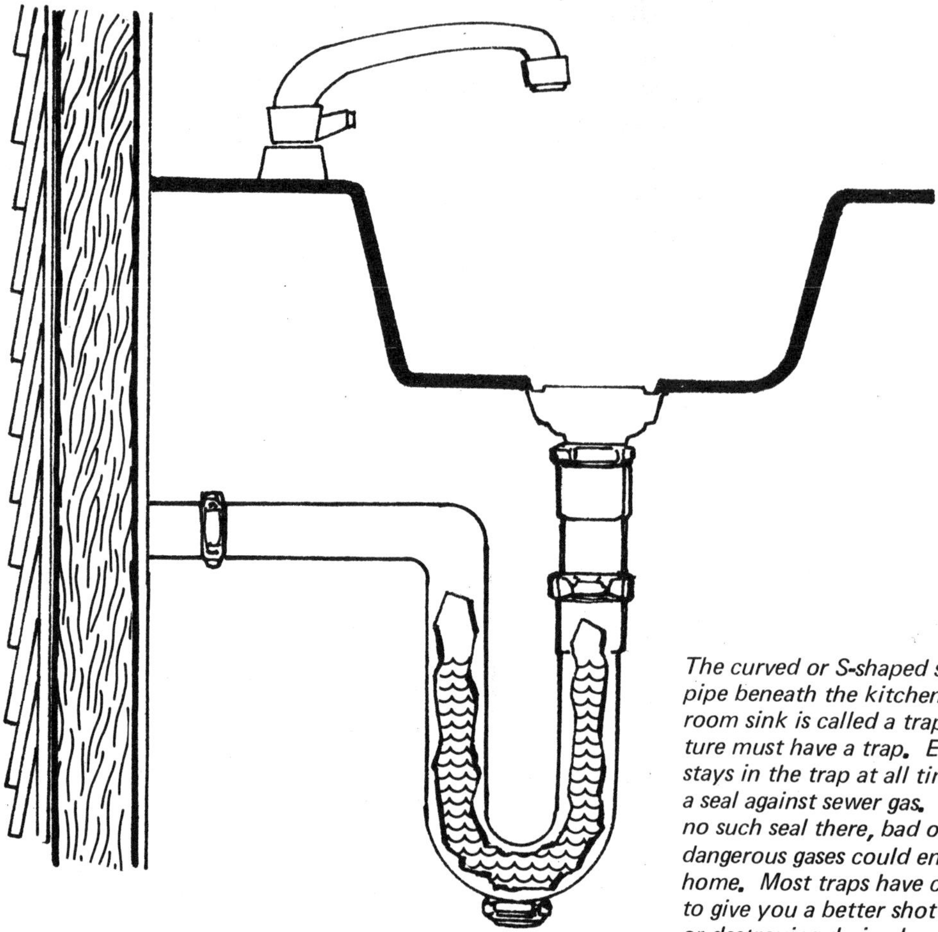

The curved or S-shaped section of pipe beneath the kitchen and bathroom sink is called a trap; every fixture must have a trap. Enough water stays in the trap at all times to form a seal against sewer gas. If there were no such seal there, bad odors and dangerous gases could enter your home. Most traps have cleanout plugs to give you a better shot at removing or destroying drain clogs.

SUPPLY AND DISPOSAL SYSTEMS

Supply	Disposal
1. Source can be the municipal water company or a private well.	1. Destination can be the municipal sewage-treatment plant or a private septic tank.
2. Water flow results from pressure — created either by a tall tank or by a water pump.	2. Waste flow results from gravitational pull.
3. Flow can be up, down or sideways.	3. Flow is always downward.
4. All fixtures should have cutoffs.	4. All fixtures must have traps.
5. Supply system must operate totally independent of the disposal system.	5. Disposal system must operate totally independent of the supply system.
6. Possible defects: leaks, drips, noises, insufficient pressure, excessive pressure, freezing, condensation, flow restrictions from scale and water chemicals.	6. Possible defects: clogs, noises, freezing, improper venting.

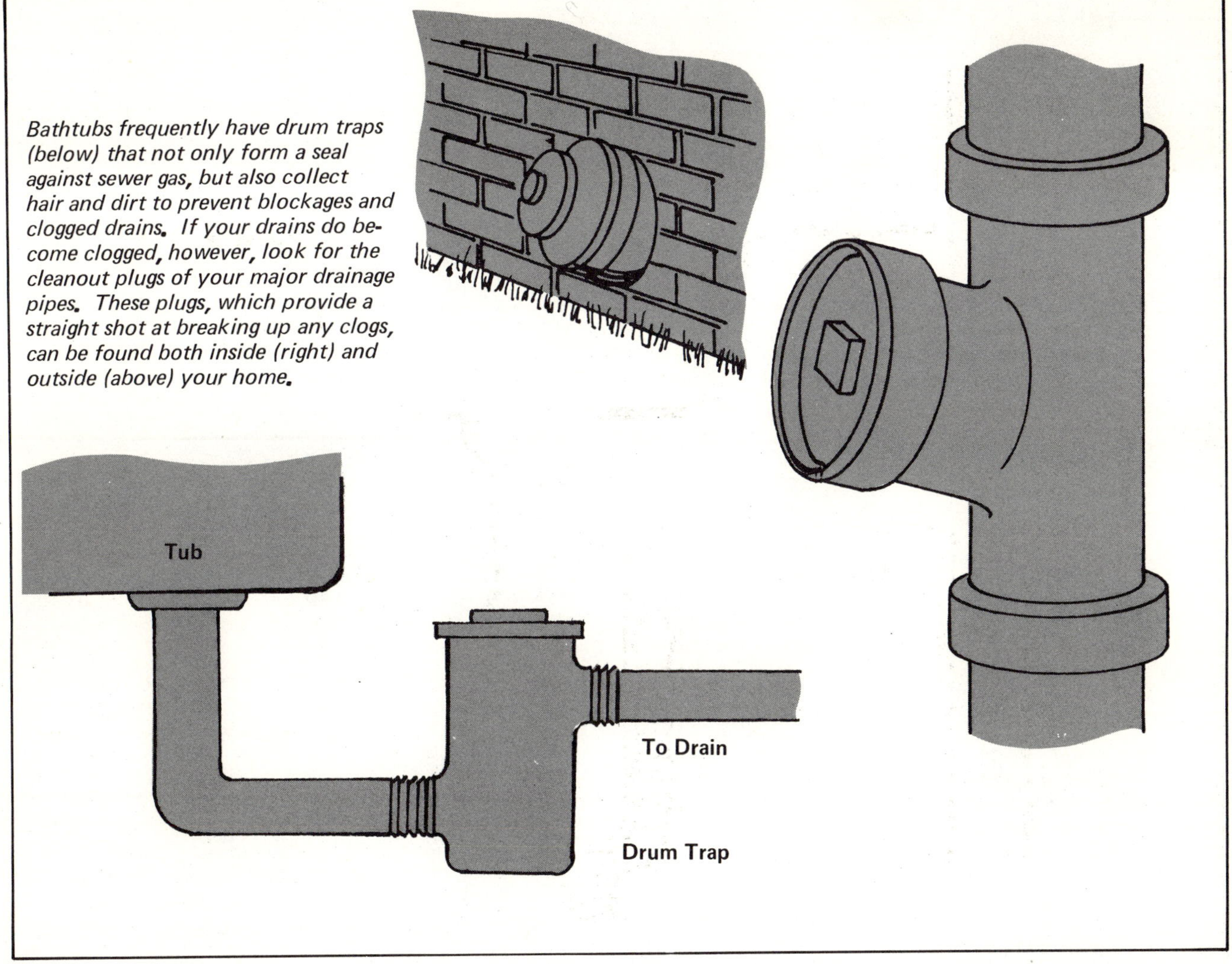

Bathtubs frequently have drum traps (below) that not only form a seal against sewer gas, but also collect hair and dirt to prevent blockages and clogged drains. If your drains do become clogged, however, look for the cleanout plugs of your major drainage pipes. These plugs, which provide a straight shot at breaking up any clogs, can be found both inside (right) and outside (above) your home.

traps have cleanout plugs that give you a direct shot at removing or breaking up any clogs.

Since the drainage system involves all of these components, it is usually referred to as the DWV — the drain, waste, vent system. If water is to flow out freely and waste to exit properly, all of the components of the DWV must be present and in good working order. Examine the pipes in your basement or crawl space. The larger and heavier pipes are for drainage. It is always a good idea to tag as many pipes as possible to know exactly what each one does. In addition, try to locate the cleanout plugs on the traps, and make sure that you know where all the vents are.

PUTTING IT ALL TOGETHER

The supply and drainage systems are — and must always be — two distinct operations with no overlapping whatsoever. There is a bridge between the two, however, and the bridge is what makes the plumbing system worth having. Plumbing jargon refers to any bridge between the supply and drainage systems as a fixture.

Toilets, sinks, tubs — these are all fixtures; but there are plenty more. An outside hydrant is a fixture and so is your new twelve-cycle automatic washing machine. All devices that draw fresh water and discharge waste water are fixtures, and all are designed to keep the supply and drainage systems they depend upon strictly segregated.

Supply, fixtures, drainage — that is all there is to plumbing. Is that anything to fear? Of course not! The fact that it is usually hidden in the walls, under the floors, and beneath the ground may make plumbing seem mysterious, but it is really a simple and logical system. Get to know how your plumbing works and you never need fear becoming one of those hapless victims that the cartoonists love to portray. In fact, once you understand plumbing, you should be able to save yourself plenty of money the next time a plumbing problem presents itself in your home or apartment.

Chapter Two

Pipes

Take a look at your pipes. How much do you know about them? Can you identify what they are made of? Do you know how to remove a bad section of pipe and install a replacement? Can you stop a leak, thaw a pipe that is frozen solid, or eliminate the racket created by pipes that bang constantly? Pipes are the lifelines of your plumbing system, and you must know all you can about how they work and how to get them back working when, for one reason or another, they fail.

PIPE TYPES

Unless you are building a new home, there is little chance that you can determine the type of pipe to be used in your house. Your choice of replacement or additional pipe in an already existing home is limited by what is there presently and by the local plumbing code. The code, of course, also dictates what can and cannot be used in new construction. Despite the fact that your choice is limited, however, you should at least know what types of pipe exist and how they are used.

Copper pipe: One type of pipe that is on the approved list in nearly every local plumbing code is rigid copper pipe. Lightweight and easy to handle, copper pipe is currently the most popular type of pipe in use. It resists corrosion and scaling; in fact, it is so durable that under normal conditions the copper pipe that you have today will probably outlast both you and your house. Moreover, the extremely smooth inside surface of copper pipe offers little resistance to water, a property which allows a copper pipe of smaller diameter to handle the same job performed by larger pipes made of some other metals.

Rigid copper pipe comes in three different thicknesses for different plumbing purposes. Type K is the heaviest version and, therefore, the type used mainly for underground installations. Type L is the medium weight copper pipe; it is used for interior plumbing. Type M is the lightest version of copper pipe, and it is also used — where local plumbing codes permit it — for interior plumbing. Even though some local codes rule out Type M copper pipe, it is adequate for most home plumbing requirements. Nevertheless, if your local plumbing code prohibits Type M, use some other weight of copper or another kind of metal pipe.

There is one more kind of rigid copper pipe. It is called DWV, and it is used only for drain, waste, and vent. DWV is even thinner than Type M copper pipe, and some plumbing codes forbid it just as they do Type M. Unlike Type M, nevertheless, DWV copper pipe does not have to withstand any pressure. Wall thickness in a supply system is very important, but it is not a crucial consideration in the drainage system. Again, however, if the local plumbing code says no to DWV, find another type of pipe.

Flexible copper tubing: You can buy flexible copper tubing only in Type K and Type L, but that disadvantage is about the only one you will encounter. Flexible copper tubing can do just about anything that rigid copper pipe can do, and plenty more besides. You can bend the tubing around corners or snake it through walls and over ceilings much easier than you can maneuver the rigid copper pipe. Its biggest advantage, though, is in the fact that flexible copper tubing comes in rolls of up to 60 feet long. When you work with rigid copper pipe you must sweat solder the two sections to be joined, and every connection is a potential source for a leak. With flexible copper tubing, however, you just unroll the length you need; the only connections you have to make are at each end. The fewer joints, of course, mean fewer chances for a leak.

Keep in mind that the larger the diameter of the tube, the easier it is to get a kink — which is the last thing you want to have happen to your copper tubing. The tube need not stay perfectly round, but a kink will result in restricted water flow. You can form gradual bends in the tubing just by using your hands; your knee is an excellent device for creating a curve. If you need to make a sharp turn, however, do not try to do it by hand; buy one of those spring-like gadgets called a tube bender that is designed specifically for such purposes.

Galvanized steel pipe: Although copper pipe and tubing have replaced galvanized steel pipe as the most popular types for home water supply lines, galvanized — which used to hold a monopoly on the supply pipe market — still offers some distinct advantages. For example, suppose your pipes run through a garage or basement or any other area where they are exposed to blows from cars or tools. For such situations you would be much smarter to use galvanized steel pipe because it is much better able to withstand damaging shocks.

The price you pay for using galvanized steel pipe is both an advantage and a disadvantage. On the one hand, galvanized is much less expensive than copper at the time of initial purchase. On the other hand, galvanized is much more expensive to repair. It takes so much longer to cut, thread, and join galvanized pipe than it does for copper pipe or tubing that labor costs will cause you plenty of regrets if you ever have to call in a plumber.

Also on the negative side of the ledger is galvanized's inability to resist corrosion from either alkaline or acidic deposits in the water. Lime and scale can build up in galvanized steel pipe, restricting the flow of water. Even at its best, galvanized pipe cannot match copper for water flow properties. The surface inside a galvanized steel pipe is not as slick as the inside surface of a copper pipe, and the fittings that accompany galvanized pipe tend to reduce the flow as well.

Plastic pipe: There is one handicap to using plastic pipe — many building codes prohibit it. Most likely, however, these codes will change in the future because plastic pipe possesses so many fine features that eventually the inertia that slows down changes in the codes will be overcome. It is difficult to say which one of the many aspects of plastic pipe is its most advantageous. It is not subject to corrosion, scaling, or rust; it is virtually self-cleaning; it will not rot, and usually it does not sweat; it can withstand freezing temperatures much better than metal can; and its weight is so light that plastic pipe is easier to handle than just about any other kind of pipe. To cap all of these fine features, plastic pipe is less expensive than even galvanized steel pipe.

The three most popular types of plastic pipe are PVC (polyvinyl chloride), ABS (acrylonitrile-butadiene-styrene), and CPVC (chlorinated polyvinyl chloride). Only the CPVC pipe is capable of handling hot water, but all three will handle cold with no difficulty. PVC, ABS, and CPVC are all rigid types of pipe; if you really need a flexible type, look for polyethylene pipe. Check your local plumbing code carefully and if the code does not prohibit plastic pipe, give it serious consideration for your next plumbing installation or repair.

Cast iron: When it comes to soil and waste pipes, the traditional favorite is cast iron. It is so very durable and practically corrosion free that it can be used for underground installations with no problems. In most home drainage systems cast iron is used for the soil and waste stacks (and for the main drains), while copper tube of the DWV type is used for the branches to the fixture drains. Cast iron pipe comes in two weights: service and extra heavy. Unless your local plumbing code tells you to do otherwise, opt for the lighter service weight version when buying cast iron pipe.

Other pipe types: When weight is a factor and underground installation is not, steel drain pipes are often used in place of cast iron ones. Much lighter than cast iron pipe, steel pipe can never be used underground because it could collapse under the weight of the dirt. Lead pipe is rarely used

PIPES

	Rigid Copper	Copper Tubing
Varieties	Types K, L, M, and DMV.	Types K and L.
Characteristics	Resists corrosion and scaling; light, durable, easy to handle, smooth interior surface.	All the good points of rigid copper, plus easy bending.
Disadvantages	Type M may be unacceptable to local code; kinks easily.	Comes only in Types K and L; can kink if bent sharply.
Uses	Supply: Type K in underground installation, Type L in interior, Type M in interior. Disposal: Type DMV from fixture to main sewer.	Supply: same as rigid copper pipe.
Fittings	Sweat soldered; flared and compression fittings.	Flared and compression fittings.
Tools	Tube cutter; hacksaw with fine-toothed blade; file; flaring tool; adjustable wrench; propane torch; emery cloth; paste flux; solder.	Tube cutter; flaring tool; adjustable wrench.

anymore except as a closet bend — the pipe that joins the toilet to the soil pipe. Septic tank systems frequently employ a perforated type of fiber pipe for releasing waste liquids into the drain field.

PIPE TOOLS

If you plan to do a fair amount of plumbing repairs yourself, then your tools become of utmost importance. Many plumbing tools, of course, can be used for several other do-it-yourself tasks around the house; but no matter how much or how little you plan to use a particular tool, it should be the best of its type that you can afford. It just makes no sense to skimp on tool quality. The price difference between superior and inferior tools is generally insignificant, but the quality difference is usually substantial. Often a good tool will last a lifetime, whereas a poor tool may not make it through the project for which you bought it. How can you tell superior from inferior? The best rule of thumb is to stick with known brands or with private-label brands from known companies.

Although the tools you elect to purchase depend upon how deeply you are involved in home repairs in general and in plumbing projects in particular, there are certain basics that any self-respecting do-it-yourself plumber should have in his tool box. For working with pipes you will need Stillson wrenches. These wrenches are generally used in pairs — one holds while the other turns. A chain pipe is made for larger pipes, and a strap wrench — which works on the same principle as the chain wrench — can be used on polished pipe without marring the finish.

Copper pipe and copper tubing necessitate your having some special tools. Since you should avoid using a hacksaw on copper, be sure to have a tube cutter in your toolbox. A small size cutter, moreover, makes it much easier to work in close quarters, and you will be amazed at how many plumbing chores are performed in cramped surroundings. You will need a propane torch for sweat soldering and a flaring tool for making flared connections at pipe joints.

Just like copper pipes, iron pipes also require some special tools. When cutting iron pipe, you should have a combination vise — one that has a section with pipe jaws — or a yoke or a chain vise, both of which are made especially for pipe work. A pipe cutter tool is easier to use than a saw, but be sure that the cutter is designed for iron pipes; the tube cutter for copper is very similar in appearance. Of course, you must be prepared to remove the burrs created by the cutting. Outside burrs will come off if you file them, but you will need a reamer to remove any burrs inside the pipe.

Although you can generally purchase the

Galvanized Steel	Plastic Pipe	Cast Iron
One type only: galvanized.	PVC, ABS, CPVC, and polyethylene.	Service and extra heavy.
Resists shocks and blows; less expensive to install than copper.	Does not rust, or corrode; self-cleaning; light; usually does not sweat or freeze.	Durable; corrosion-free.
Difficult and expensive to repair; corrodes from acid or alkaline; subject to lime and scale deposits; inferior water flow.	Unacceptable to many local plumbing codes; most types cannot handle hot water.	Cutting and connecting are difficult and time-consuming procedures.
Supply: in areas where pipes subject to damage.	Cold water supply: Type CPVC can handle hot water.	Disposal: especially underground installations.
Threaded joints.	Solvent welded; attaches to copper or galvanized with adapters.	Bell-and-spigot, gasket, and no-hub connections.
Pipe cutter; hacksaw; cutting oil; pipe die; pipe wrenches.	Fine-toothed saw; miter box; solvent cement.	Hacksaw; chisel; hammer; oakum; lead; joint runner; pipe wrenches.

Pipe Tools

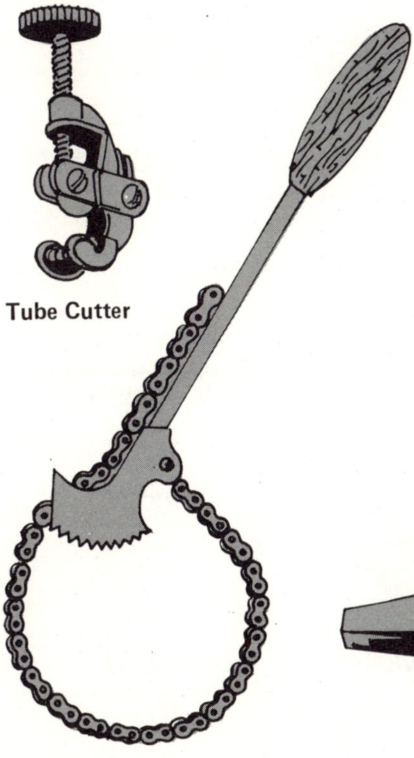

Tube Cutter

Chain Wrench

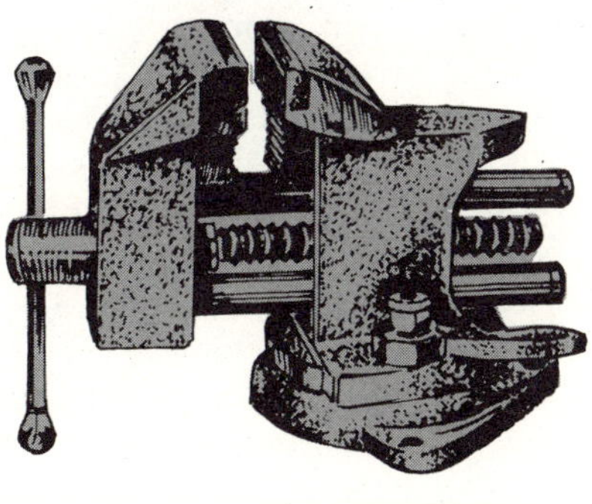

Vise With Rounded Pipe Jaws

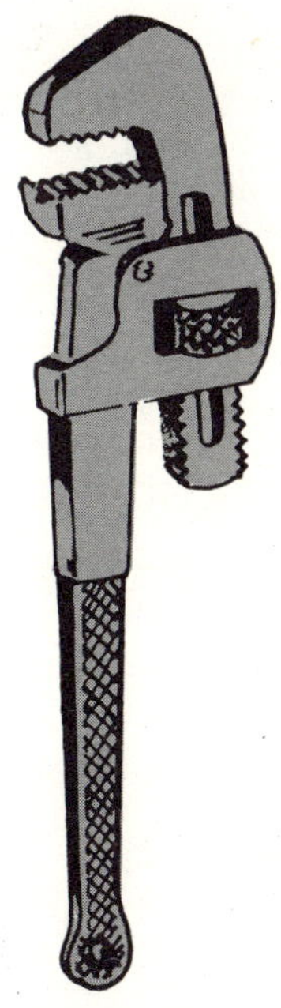

Adjustable Pipe Wrench

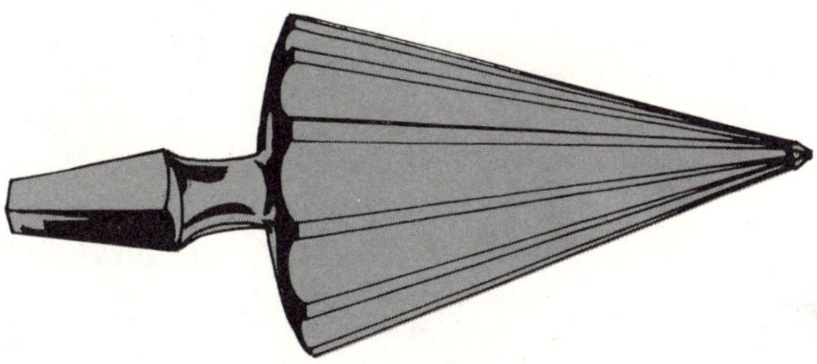

Reamer

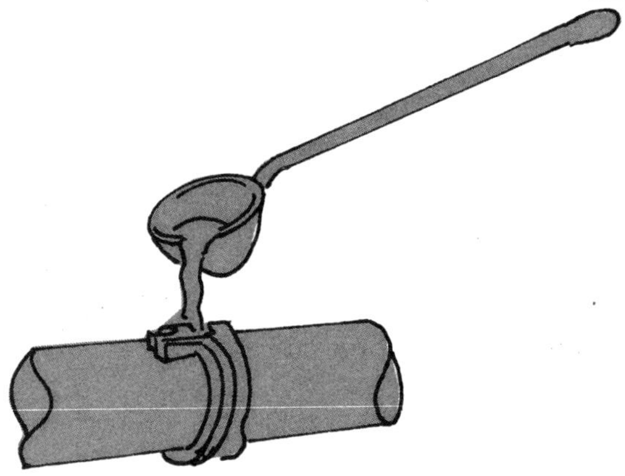

You need a special tool called an asbestos joint runner when you pour molten lead into a horizontal joint between cast iron drain pipes.

iron pipe you need already threaded, you may wish to do your own threading. If so, you will need special dies — called NPT dies — that come in sizes for threading all standard pipe diameters. Remember that whenever you are cutting or threading iron pipe you should use a cutting oil. And, speaking of oil, a can of penetrating oil always comes in handy when you tackle a plumbing repair job.

For sealing large cast iron drain pipes against leaks, you need a caulking tool for poking the oakum into the joint, a melting pot for heating the lead, and a ladle for pouring the molten lead over the oakum. When you work on horizontal drain pipes, moreover, you face the problem of keeping the molten lead in the joint and preventing it from running out on the ground. The answer to this problem is a special tool called an asbestos joint runner.

CUTTING AND CONNECTING COPPER PIPE

Cutting copper pipe or tubing without kinking it takes care. Whenever possible, avoid sawing copper; use a tube cutter instead. Of course, a tube cutter will be of little use on the larger sized copper pipes. For those you must use a hacksaw, but be sure to equip your hacksaw with the finest-toothed blade you can find before you start cutting. Saw as straight as possible, and eliminate any irregularities in the edge — as well as any burrs both inside and out — with a file. If you find that you must put the copper pipe in a vise to hold it while you saw, clamp the vise far away from where you are cutting so that there are no dents to ruin the end of the tubing or pipe. Anything but a perfectly round end will not connect well to another pipe or tube.

Copper pipes or tubes can be joined in several ways. Sweat soldering is the most common and least expensive way of joining copper pipe, while flared connections and compression fittings are used mainly for flexible copper tubing. The advantages to a flared fitting are that: (1) it can be taken apart easily with a pair of open-end wrenches; and (2) it can be used in situations where sweat soldering is either dangerous (fire hazard) or impossible (uncontrollably wet conditions). The disadvantage is that flared fittings are much more expensive than the ones you sweat solder.

To make a flared connection you need a flaring tool and a special fitting that includes a flange nut. You slip the flange nut over the perfectly square cut copper tubing, making sure that the open end of the nut faces out. Then, you insert the tubing into the proper sized hole in the flaring tool, and turn the tool down to flare the end of the tube. Inspect the flared end to make sure that it is even, that there are no burrs, and that the tube's surface is completely clean.

Put the fitting into the flared end of the tube,

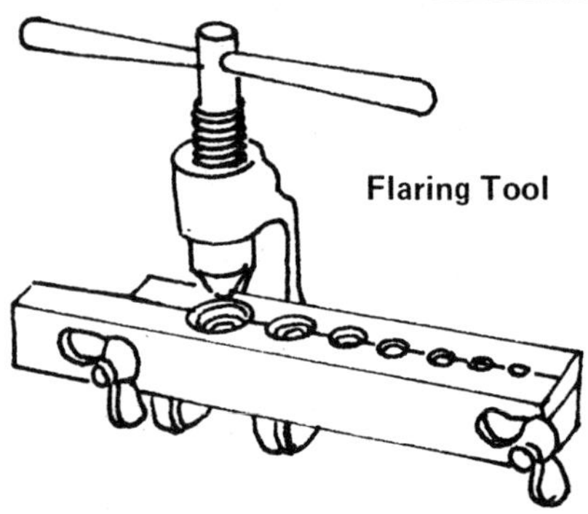

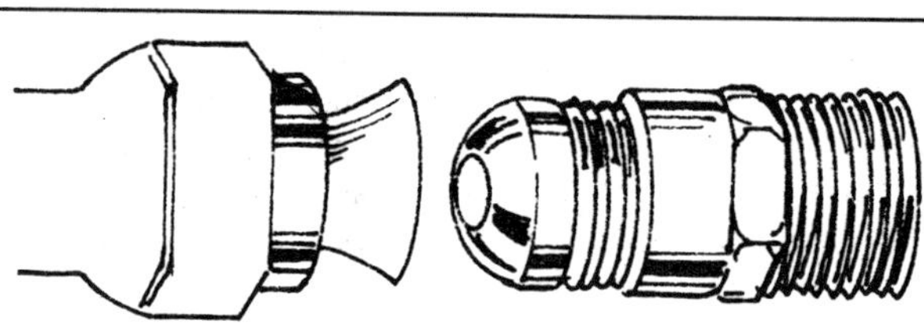

Flared fittings (top, right) and compression fittings (right) are used mainly for connecting sections of flexible copper tubing. To make a flared connection, however, you must have a flaring tool (above).

pull the flange nut up to it, and start turning the nut down tightly. When you get the flange nut as tight as you can get it by hand, use a pair of open-end wrenches — one to hold the fitting and the other to turn the flange nut — to complete the flared connection.

The compression fitting offers the same advantages — and the same price disadvantage in relation to sweat soldering — as does the flared fitting, but it does not require a flaring tool. It has the same two parts as the flared fitting, but the compression fitting also has a compression ring. You slip the flange nut onto the tubing (again, the end of the tube must be cut perfectly square), slip the ring over the tube as well, and then insert the tube into the fitting as far as it will go. Hand tighten the flange nut, and then use two open-end wrenches — one to turn the nut and the other to hold the fitting — to seal the ring against the fitting and to squeeze it against the tube.

SWEAT SOLDERING

Sweat soldering is the most popular method for connecting copper pipes. Done properly, sweat soldering is both effective and inexpensive. If not done right, however, sweat soldering certainly is not effective — and it can be disastrously expensive.

Unlike conventional soldering, you sweat solder with a torch rather than an iron or a gun. Most people now use a propane torch, which is easier and safer to operate than a blow torch. In addition to the heat source you need paste flux, emery cloth, and solid-core solder.

At least as important as having the right materials is preparing the pipe properly. There are three key ingredients to assuring good sweat soldered joints: (1) clean copper; (2) totally dry pipes; and (3) proper heat. If you are dealing with newly cut pipes or tubing, be sure to remove all the burrs from both inside and out. Of course, if there are any indications of the tube having been bent or crushed in the cutting process, start over with a freshly cut piece of copper. Then, check the fittings to be sure that the sweat soldered joint will be a tight one.

Now, clean all the metal surfaces that you plan to join. Fine steel wool or fine sandpaper are adequate for the job, but the best way to clean the copper is to wrap emery cloth strips around the tubing or pipe and go back and forth as if you were shining a pair of shoes. You want the metal to shine, but be careful not to overdo it. The last thing you want to do is to take away so much metal that you destroy a tight fit. On the other hand, do not forget to clean every surface that will be involved in the sweat soldered joint, including inside the fitting where the pipe is supposed to go.

When you are through with the emery cloth strips, coat all the cleaned metal surfaces with flux, which is a paste especially compounded for copper soldering. Spread the flux on both the pipe and the fitting, and — after you insert the pipe in the fitting — rotate them a bit to distribute the paste around the joint.

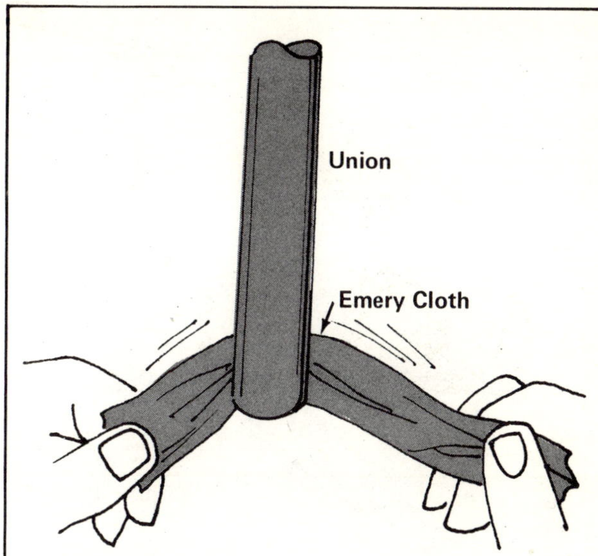

The best way to clean the surface of rigid copper pipe prior to sweat soldering is to wrap emery cloth strips around the pipe and then go back and forth as if you were shining a pair of shoes. Although you must clean the metal thoroughly, be careful not to take away so much of the pipe that you destroy a tight fit.

One of the most crucial factors in sweat soldering is making sure that all the pipes are completely dry. If there is any residual moisture, it will turn to steam when you start heating the pipes, and the end result will be a weak or even a leaking joint. Therefore, before you actually start putting the pieces together, make absolutely certain that no moisture is lurking anywhere. There are a couple of techniques that will help you to dry the pipes thoroughly. If you are sweat soldering to a vertical pipe, you can use a rag to remove any moisture; but if the connection is to be made to part of a horizontal run, try this trick of the plumbing trade. Wad up a piece of bread into a ball, poke the ball into the pipe, and push it back beyond the area you will be soldering. The bread will absorb any drops of moisture that might otherwise reach the joint and ruin your sweat soldering. When the job is done and you turn the water back on, the bread will dissolve and the residue will come out through the nearest open tap.

Once your copper pipe is completely clean and totally dry — and you have applied the flux and joined the pipe and the fitting — you are ready to perform the actual soldering. Light your propane torch and get the solid-core solder ready in your other hand. Play the flame over the fitting, overlapping the edge and onto the fitting. Do not put the flame on the solder; the solder is melted when it is

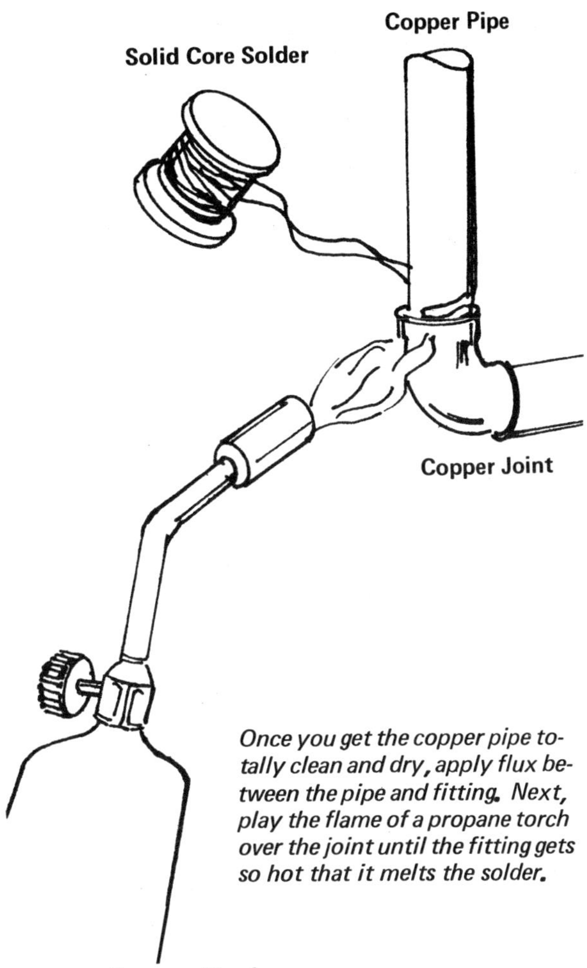

Once you get the copper pipe totally clean and dry, apply flux between the pipe and fitting. Next, play the flame of a propane torch over the joint until the fitting gets so hot that it melts the solder.

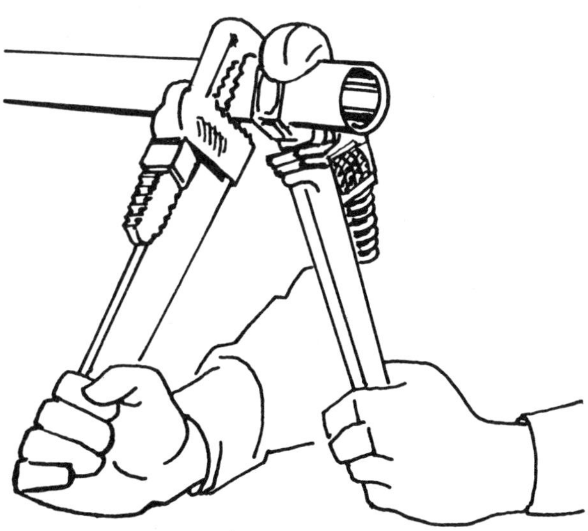

Fittings for galvanized steel pipe are threaded, as are the ends of the pipe sections. To connect them, grip the fitting with one wrench and the pipe with another and tighten until only three threads still show.

brought into contact with the hot fitting and pipe.

As you continue to play the flame over the fitting, bring the end of the solder into contact with the lip of the fitting. When the metal gets hot enough to melt the solder, you will witness an amazing law of nature called capillary action. Not only does the solder melt, but it also gets sucked up into the joint under the fitting; this capillary action occurs even when the pipe runs vertically up into the fitting. Keep heating the fitting and melting the solder until the entire joint is filled; you know that point is achieved when you see a bead of solder form around the lip. For neater joints, you can wipe away any excess solder by taking a quick swipe around the lip with a piece of coarse toweling, but be careful not to touch the hot metal with your hand.

The entire process is actually quite simple, but it does demand that you keep some safety precautions clearly in mind. If you are working close to a wall or ceiling, place a scrap of asbestos or a piece of aluminum foil between the pipe (or tube) and the wall or ceiling. Otherwise, were you accidentally to point the torch in the wrong direction you could char your surroundings or even start a fire. Be especially careful, of course, to stay away from already soldered joints or valves nearby. Wrapping wet rags around good joints and valves is a good way of dissipating any extraneous heat and preventing the flame from doing any damage. One other safety rule: when you finish sweat soldering the joint, turn off the torch immediately.

Many amateur plumbers are so proud of their first sweat soldered joints that they run to turn on the water. That, however, is a big mistake. Be sure to allow the joint to cool naturally; the sudden cooling effect of the rushing water can weaken the joint or even cause it to crack. If, when you finally do turn on the water, you discover a leak in the sweat soldered joint, start the whole job over from scratch: cut off the water, melt the solder, remove the fitting, dry and reclean the fitting, reflux, and finally resolder. There are instances when you can get away without starting over, but it is generally easier, quicker, and more effective to sweat solder the joint again as though you had not done it before.

Good sweat solderers gain their expertise through practice. Therefore, why not buy a few short lengths of copper pipe and some fittings now, and try your hand at it before you are faced with a genuine sweat soldering situation? After you do a few joints, you will be sweat soldering like a pro.

WORKING WITH OTHER KINDS OF PIPE

Galvanized pipe is usually cut with a hacksaw, although a pipe cutter — if you have access to one — makes the job much easier. Be sure to use cutting oil and to place the pipe in a vise while you cut it.

Since galvanized steel pipe is connected

with threaded joints, you must thread the new end after each cut. You thread galvanized pipe with a pipe die. When you are ready to join the pipe to its fitting, smear the threads with pipe joint compound to help seal the joint. Teflon tape wrapped around the threads of the galvanized pipe and pressed down into the grooves makes the joint watertight.

The fittings for galvanized pipe should be hand tightened first, and then tightened securely with a pair of pipe wrenches. Turn until there are only about three threads still showing outside the fitting. In addition to the standard fittings for galvanized pipe, there are short lengths of pipe available (called nipples) that are threaded on both ends. Nipples come in lengths of from one to twelve inches.

Although cutting and connecting galvanized pipe is not difficult, it cannot match the ease and convenience of working with plastic pipe. Plastic pipe is a joy to cut, and joining two sections is almost as simple as gluing photos in an album. You can cut plastic pipe with any fine-toothed saw; a hacksaw is as good as any and better than most. Sawing plastic requires very little effort, and light sanding can get rid of all the little burrs both inside and out. Of course, plastic pipe — like any other type — must be cut square. To get a square cut, place the plastic pipe in the right-angle slots of an inexpensive wooden miter box; then start cutting.

Solvent welding is the name of the process used to join sections of plastic pipe. All it involves is brushing a special solvent cement on the end of the pipe and on the inside of the fitting, and then putting the pipe and the fitting together. Give the two a quick twist to spread the solvent, but then get them positioned correctly without hesitating because the solvent goes right to work in forming the bond. It actually fuses the two parts together. Hold the fitting in place for about fifteen seconds to prevent any slippage. After that, the weld is set.

There are certain types of plastic pipe installations that must be made somewhat differently. With CPVC pipe, for example, you use a two-phase solvent weld system; the first part is the application of a type of flux. When joining plastic pipe to other types of pipe, you are faced with yet another procedure. Suppose you want to use plastic pipe for an underground sprinkler system. At some point, you must attach the plastic to your existing plumbing, which would be either copper or galvanized. You can make the connection easily with the adapters designed specifically for linking plastic with other pipe types.

Far more difficult to work with than either galvanized or plastic is cast iron pipe. Smart do-it-yourself plumbers always plan ahead when it comes to dealing with cast iron; they try to arrange for as few cuts as possible. Probably the best way to make the cut is to draw a line around the pipe with chalk, and then to score the pipe with a hacksaw all the way around at the chalk mark. Be sure that the cut is

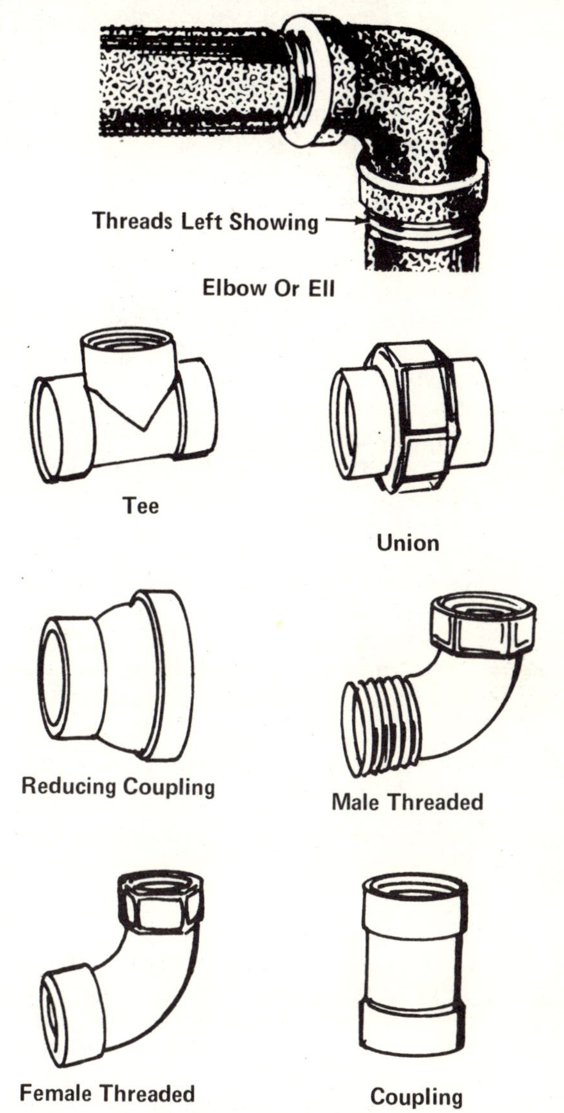

Threaded fittings come in many different styles so that you can run your pipes in nearly any direction and around any sort of obstruction in their path.

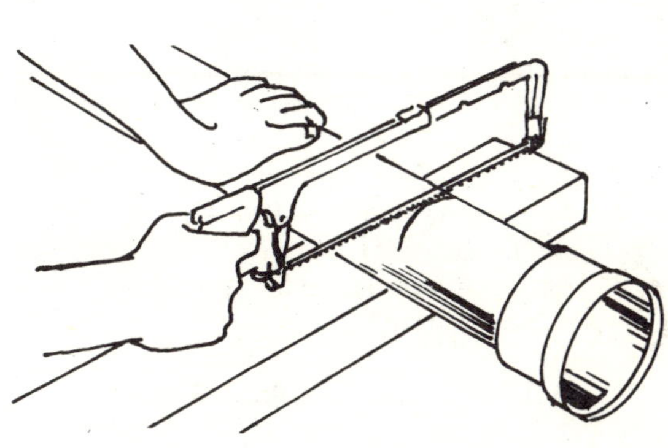

To cut cast iron pipes, use a hacksaw to score a line all the way around the circumference of the pipe. Then tap the end of the pipe with a hammer until it breaks off.

square and that you score the pipe to a depth of about 1/16 of an inch. Then, if you are cutting the lighter service weight pipe, place it on a scrap block of wood with the scored line overhanging slightly and start tapping with a hammer on the part that is supposed to come off. Eventually, the pipe should break cleanly. If you are using the heavier cast iron pipe, cut on through the scored line with a cold chisel.

Cast iron pipe can be joined in any of three ways: the bell-and-spigot joint, the gasket-type connection, and the no-hub system. The bell-and-spigot is the oldest of the three, but it is still in widespread use. In this type of joint one end of the pipe has a bell-shaped hub into which the slightly ridged end of the connecting pipe fits. You will find that there is space to spare between the two sections of pipe; you fill this space with a sort of oily rope called oakum. Wrap the oakum around the pipe, and then poke it down into the gap between the two sections of pipe.

When the oakum fills the void up to about an inch (different plumbing codes specify different depths), you pour molten lead over it. As soon as the lead starts to harden, you work it down tight against the oakum and the sides of the pipe. You follow the same procedure whether you are connecting two sections of cast iron pipe or a pipe to a fitting. The only time you do anything differently is when you are working on a horizontal joint. Then you do everything just as you would for a vertical joint, except that you pour in the molten lead with a special gadget called a joint runner.

Gasket-type connections also include a belled hub on one pipe, but the other pipe end is hubless. A neoprene gasket inserted into the bell holds the connection together. When you force the plain end of the hubless pipe into the hub (with the aid of a lubricant), the result is a watertight joint.

The newest and easiest way to join cast iron pipe is called the no-hub system. In this system the ends of all the pipes and the fittings are plain; they are joined by a neoprene sleeve that is held in place by a shield. The shield, in turn, is kept tight by clamps which are similar to worm-gear hose clamps. No-hub connections are easy to put together, and they have the definite advantage over other cast iron connections in that they are easy to take apart. Their only disadvantage — the same one that plagues plastic pipe — is that some local plumbing codes place no-hub connections on the forbidden list; many codes specifically exclude them from underground installations. Where a no-hub connection is allowed, however, smart plumbers use it.

ADDING OR REPLACING PIPE

Installing new plumbing is never a pipe dream, but it need not be a nightmare either. There are certain basic ground rules and many useful techniques that any do-it-yourselfer should know before adding or replacing any kind of pipe.

Remember that the drainage lines work on the basis of gravitational pull, not water pressure. Therefore, when you are adding or replacing drain pipes, be sure that they are pitched so that the downward flow will carry out the waste. The normal degree of pitch is a quarter inch per foot.

All runs of pipe must be supported properly — either by hangers, straps, notches in the studs or joints, or by whatever method is best for the given situation. If you notch a joist, be sure to compensate for the lost strength by nailing a brace strip under the pipe; notches in studs should be reinforced with metal mending plates. It is always best to run the pipe across the top half of the studs because the higher the notches the less the studs are weakened. Moreover, drilling a hole for a pipe run weakens a

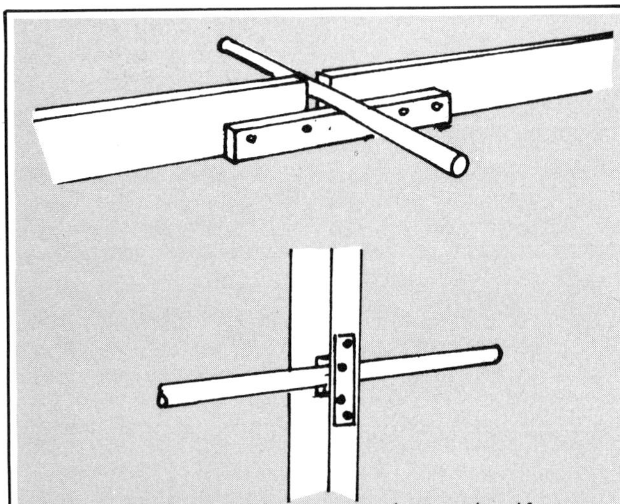

All runs of pipe must be supported properly. If you notch a joist (top), nail a brace strip under the pipe; if you notch a stud (bottom), add a metal mending plate.

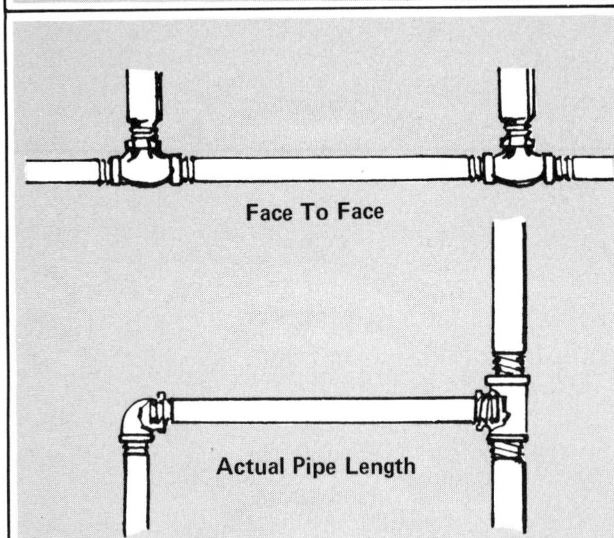

To measure pipe properly, you must take into account the amount added by the fitting, as well as the amount of each pipe that goes into the fitting.

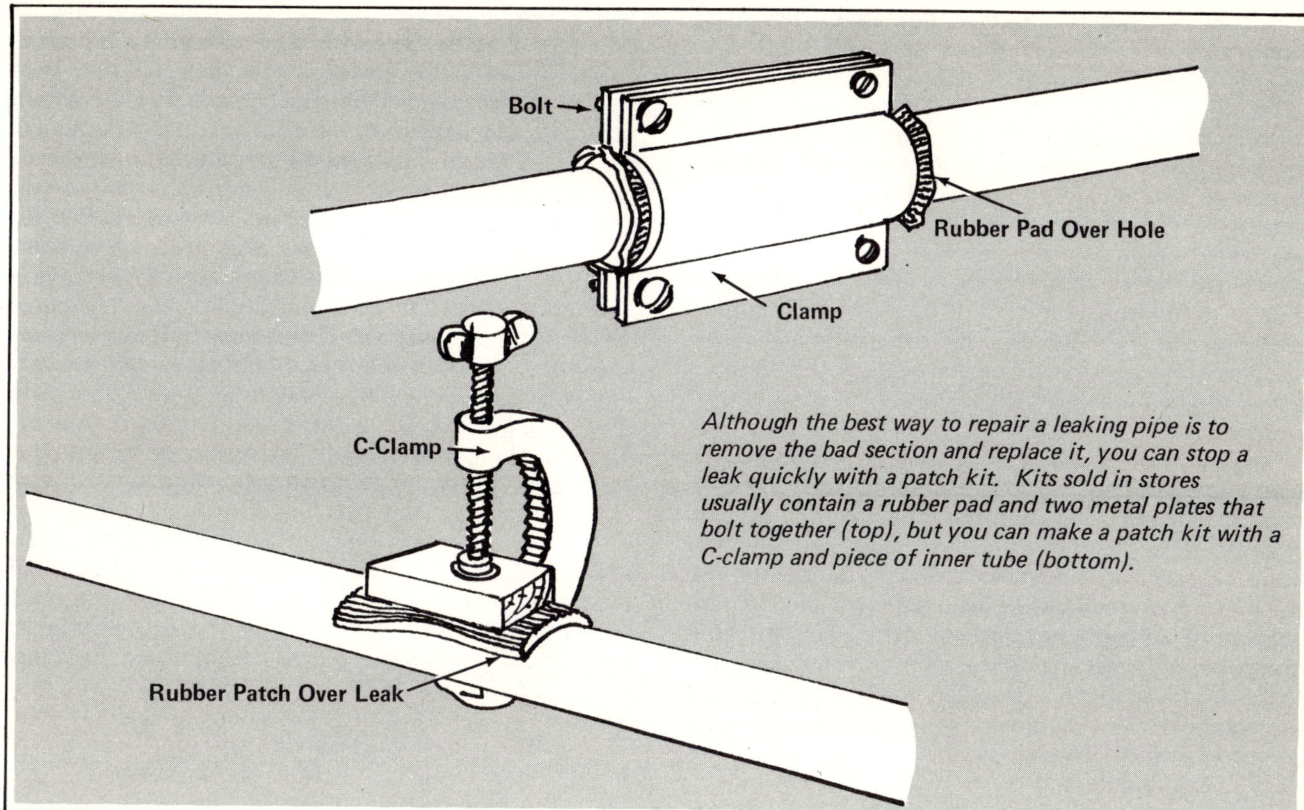

Although the best way to repair a leaking pipe is to remove the bad section and replace it, you can stop a leak quickly with a patch kit. Kits sold in stores usually contain a rubber pad and two metal plates that bolt together (top), but you can make a patch kit with a C-clamp and piece of inner tube (bottom).

joist or stud less than cutting a notch does.

Plan the entire run of pipe carefully before you start doing any work whatsoever. If you must tear into floors or walls to correct pipe problems, consider substituting a new line over an entirely different course. Naturally, there are questions of aesthetics, but even running pipe along the outside of an interior wall may — when properly boxed — be preferable to ripping into the walls themselves. Frequently, you can hide a visible stretch of pipe by running it through a closet.

Once you decide exactly how you plan to install your new or replacement pipe, you must know how to order the lengths you will need. To measure pipe properly you must take into account two factors over and above the length you can see: (1) "fitting gain," or the amount added by the fitting; and (2) "make-up," the amount of each end of the pipe that goes into the fitting. Since pipes in most fittings do not butt up against each other, the fitting gain must be added to the length for the overall run and the make-up must be added to the individual sections of pipe. Threaded fittings are so standardized that your hardware or plumbing supply store can show you a table that tells you the screw-in distance; then you will have no problem in measuring the pipes and fittings you plan to install.

With most copper fittings, you can see from the outside just how much of the copper pipe goes inside, and with flexible copper tubing precise measuring is not critical because you can generally make the tubing fit — provided you have sufficient length — with a little gentle bending. Bell-and-spigot connections for cast iron pipe, however, require that you measure the distance from the bell lip down to the place where the spigot will rest. Since the no-hub fittings butt up against each other, there is no gain nor make-up to consider. Connections or joints should always be kept to a minimum in every run of pipe since each joint represents a possible source for a leak.

Most importantly, before you add or change any pipe anywhere in your house, check the plumbing code for your community to learn what is allowed and what is forbidden.

STOPPING LEAKS IN PIPES AND JOINTS

There are all kinds of leaks; some can flood your home, while others are not nearly so damaging. Your approach to stopping a particular leak, therefore, depends on the type of leak it happens to be.

The safest and surest way to repair a leaking pipe is to remove the section that has the leak in it and replace it with a new section. But that is more easily said than done. When you turn a pipe to unscrew it from its fitting at one end, you tighten the pipe into its fitting at the other end. There is, of course, a solution to this dilemma; but before going into that, consider another alternative: the pipe patch.

You will find patch kits at your local hardware store, or you can rig up your own with a piece of rubber from an old inner tube and a C-clamp. The factory-made kits generally contain a rubber pad that goes over the hole in the pipe and metal plates that bolt tightly together to compress the rubber pad against the hole in the pipe. The homemade rubber patch and the C-clamp do the same thing, but you may wish to place a block of wood against the pad so as to spread the clamping pressure better. A quick and easy way to stop a leak, the patch kit can even be used on a permanent basis if the pipe is otherwise sound.

Other quick and easy measures for stopping pipe leaks include wrapping waterproof tape over the bad spot, rubbing the hole with a stick of special compound, or inserting a self-tapping plug into the hole. When using the waterproof tape, be sure to dry the pipe thoroughly before you start wrapping. Then, start the tape about two or three inches from the hole and extend it the same distance beyond. For tiny leaks in pipes, there is probably no easier cure than the compound sticks — such as Krak-Stik — that are available at most hardware stores. You just rub the stick over the hole and the leak stops; Krak-Stik can even stop small leaks while the water is still running in the pipe. The self-tapping plugs should be used only for leaks in large pipes or for leaks in tanks; they would impede the flow of water in smaller pipes. When tightened down, the screw applies enough pressure to stop the leak.

The problem with these solutions is that a pipe which is bad enough to spring one leak, often starts leaking in other locations as well. You may fix one spot, only to see the pipe burst somewhere else. Especially in cases where the leak results from substantial corrosion, the chances are that the whole section of pipe needs replacing.

That gets us back to the dilemma of removing the pipe that gets tighter at one end when you loosen the other. The solution to this seemingly impossible situation is to cut the pipe in the middle. You must, however, be exceedingly careful in sawing and removing so as not to put undue strain on other parts of the pipe system. Therefore, before you start sawing, make sure that the pipe is properly braced; the two stubs will be heavy, and unless you have them propped up they will sag and pull on the joints. The last thing you want to have happen is for your sawing to break the seal at an otherwise perfect joint in your plumbing system.

Remember to do one other thing before you start cutting: turn off the water. In addition, drain all the water out of the system by opening the taps below the pipe or by opening the drain valve that is built into the system. Use a fine-toothed hacksaw blade for cutting, and saw right straight through. Now, by gripping the fitting with one wrench and the pipe with another, you can turn the pipe stub out of its fittings. Once you remove both stubs you are ready to install the new pipe sections.

You replace the leaky pipe with two sections of good pipe, which are connected by a fitting called a union. A union fitting has two female threaded sections that are pulled together by turning a special nut in the middle. A union allows you to fasten the pipes together without having to twist the pipes themselves. Measure the sections of pipe

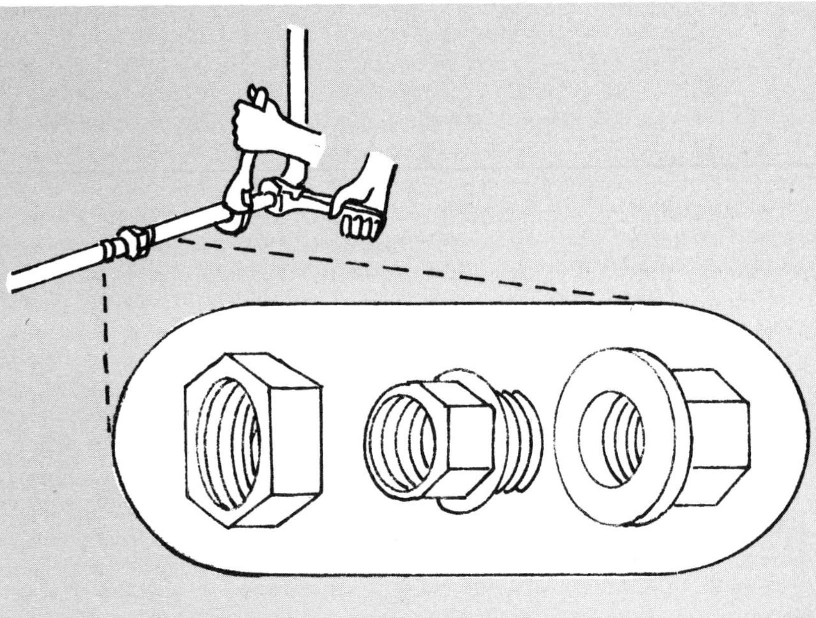

A union fitting, which possesses two female threaded sections that are pulled together when you turn a special nut in the middle, allows you to fasten two sections of replacement pipe to each other without having to twist the pipes themselves. Be sure to add in the length of the union fitting when figuring the overall run of pipe to assure a proper fit. Flexible copper tubing does not require a union.

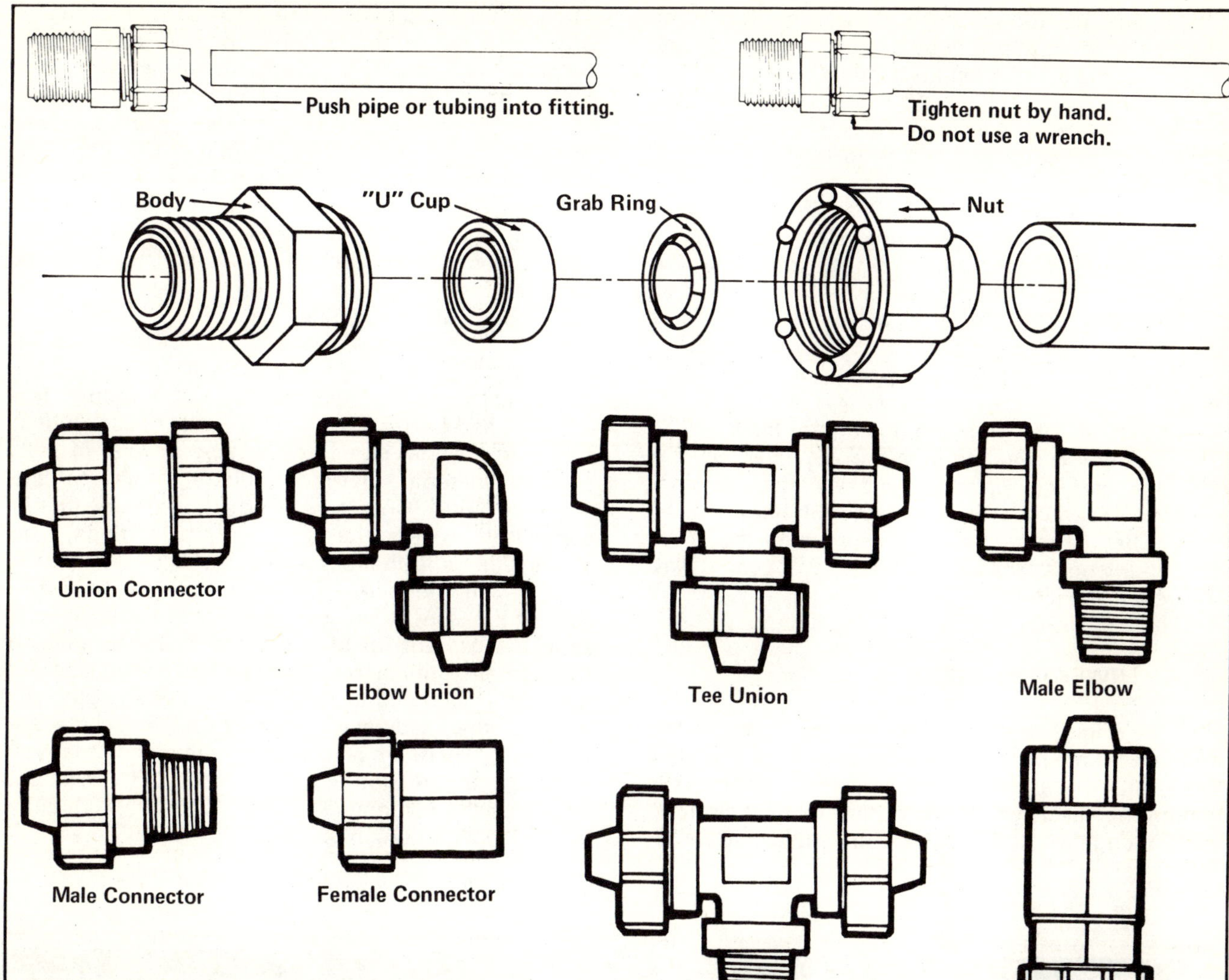

Click & Seal fittings come in eight popular versions, making them suitable for just about any situation. When water flows into the fitting, it fills the U cup and presses the cup against the tubing.

carefully to assure a proper fit, and be sure to add in the length of the union when figuring the overall run. The two new pieces of pipe must be threaded on both ends, and — as is true of all threaded metal connections — the ends should be coated with pipe joint compound or Teflon tape to make the bond watertight.

When you are replacing a section of copper pipe, you can use a tube cutter to cut through the old pipe. Then, you solder in a copper-to-copper union as well as a coupling to connect the new section. Flexible copper tubing, on the other hand, usually does not require a union. You can usually just bend the ends of the tubing slightly to get them into the fittings.

What do you do if there is a leak in a joint? That depends on what type of joint it is. If it is a leak in a sweat soldered joint in copper pipe or tubing, then you should take the joint apart and sweat solder it all over again. Leaky joints in plastic pipe usually must be cut out and replaced because the solvent that fuses pieces of plastic pipe together does such a good job that there is no way to get the pieces apart.

A threaded joint in galvanized or cast iron pipe may only need retightening when it starts to leak. Should the retightening not stop the leak, however, try applying epoxy metal to the joint. Cut off the water, dry the pipe thoroughly, and then follow the directions for applying the epoxy; that should stop the leak. Pipe joint compound can also stop threaded fittings from leaking. Merely coat the threads with the compound.

Perhaps the best and easiest way to solve a leaking joint problem is to install a new type of fitting called a Click & Seal. Made of a special Dupont nylon, Click & Seal fittings require no solvents, no solder, and no tools. Since they come in eight popular versions, they can be used in just about any situation: on plastic, copper, or other types of metal pipe, and for either hot or cold water. To be sure you buy the proper size, match the Click & Seal fitting to the outside diameter of the pipe to which it will be attached.

Look inside a Click & Seal fitting; you see only four parts: the nut, grab ring, U cup, and body. When you push a pipe into the fitting, it goes all the way into the cup. Then, the sharply angled surface of the lock ring holds the pipe or tubing in position while you hand-tighten the nut. When water flows into the fitting, it fills the U cup and presses the cup against the tubing. That is the secret of the Click & Seal: the higher the pressure, the tighter the seal.

Click & Seal fittings are approved by the FDA for carrying potables and by the National Sanitation Foundation for use with both hot and cold water. They certainly have the approval of the do-it-yourselfer who is looking for an easy and secure way to cure leaking joints.

SWEATING PIPES

Sometimes you can find so much water dripping from a pipe that you become convinced that there must be a leak somewhere. On closer examination, however, you discover that there is no leak, even though your floors and furnishings suffer as much damage as though there were. What you are witnessing is a natural phenomenon called condensation or "sweating." Sweating occurs when the water inside the pipe is much colder than the surrounding air, a situation that exists nearly the year round in many homes. During summer the surrounding air is naturally hot, and during the winter the air is heated by the furnace. In either case, when the warm air reaches the cold pipes (and the water which comes from below ground is always fairly cool), drops of moisture form and drip down as if there were a tiny hole in the pipe.

There are two ways to control the moisture problem of a sweating pipe. One is to place pans strategically on the floor to catch the drops. The other — and better way — is to insulate the pipes. There are several types of drip tape in rolls of self-sticking material that can adhere easily to the problem pipes. Once you wind the tape so that it completely covers the pipe and the fittings, you should see no further signs of sweating.

Another sweat fighter is asbestos tape, but you must use it in conjunction with a special asbestos cement for best results. Soak the tape in the cement until it becomes like papier-mache for molding around odd-shaped fittings. For straight runs of pipe, the ready-made asbestos pipe jackets are excellent. There are also no-drip compounds that you can brush on the pipe to form a coating of insulation. Although the compound is easy to apply, frequently you will have to brush on a second coat.

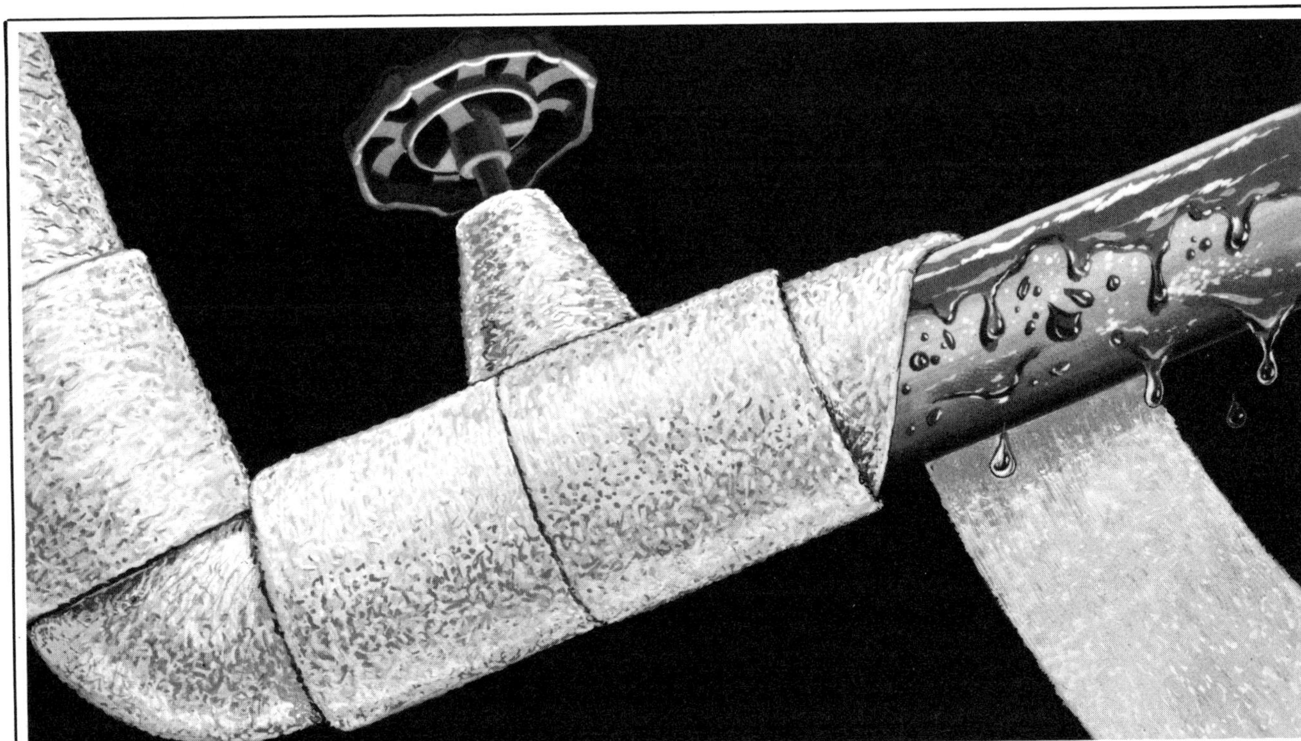

The best method for controlling pipes that sweat is to wrap them with an insulating tape like Mortell's No Drip.

It is well worth your time, effort, and expense to eliminate condensation problems. The moisture that drips from sweating pipes can damage your floors, encourage mildew, and do a host of other destructive things to your home.

DISASTER LEAKS

Leaks that produce puddles are one thing, but leaks that can flood your home with thousands of gallons of water are another story altogether. The pressure of the water supply system is so strong that by the time you can call a plumber, a leaking pipe can do thousands of dollars in damage to your house and its furnishings.

Plumbing emergencies need not become plumbing disasters, however. If you take the time now — when there is no water rising in your basement — you can learn how to prevent personal injuries and to limit property damage should a pipe ever spring a serious leak. The first thing to do is to make sure that everyone in the household knows how to shut off the flow of water. The second thing to do is to make sure that everyone knows how to shut off the flow of electric current.

Usually located very close to where the water supply enters your house, the main water cutoff may be a gate valve or it may be an "L"-shaped rod. If the water meter is in your basement, you could have a valve cutoff that requires a wrench for turning. In any case, find out exactly how the main water cutoff works and — most importantly — if it works. Little-used valves have a way of corroding, and it is much better to find out now that you need a new valve than when you are in a panic situation. As for preventive maintenance, it is always a good idea to turn the water cutoff valve off and then on again once every six months or so to keep it in working order. Another smart idea is to tag the cutoff with a drawing that indicates which way you must twist it to turn it off.

It is equally important that everyone in the family know how to throw the master switch that cuts off all the electric current throughout the household. Remember that the flooding waters in your house can come in contact with electrical components. Make sure that you do not complete the fatal triangle. Any time there is even a remote chance of a water-electricity contact, do not start wading around in the water until you throw the main switch to your household current.

With the water and power lines off, you can take your time and calmly evaluate the damage. You can determine whether the pipe that burst is otherwise sound and can function well after patching, or whether you must replace the faulty — and possibly corroded — section with new pipe. If you think that patching will do the job but do not want to use the rubber pad and clamp kits, consider applying epoxy metal for a permanent patching job. Just be sure to follow the directions and allow for the full curing time.

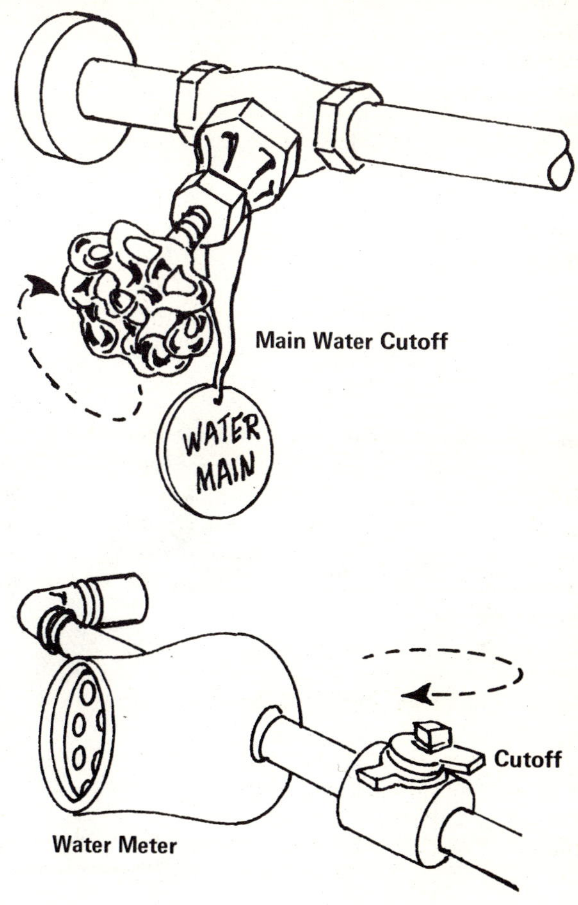

You can prevent disaster leaks from damaging your home and its funishings by shutting off the flow of water at the main cutoff valve. Be sure you know how to operate the valve; some can be turned by hand (top), while others require a wrench (bottom).

A break in a supply line certainly constitutes the worst sort of plumbing emergency, but some people regard an overflowing clogged drain as an emergency situation as well. To take the clogged drain out of the emergency category and into the simple repair category, however, all you need do is to make sure that no additional water is poured in the drain until you can clear it. This, nevertheless, can often be a tricky assignment. Frequently, one drain serves several outlets, and you must make certain that no water flows into any of them until you remove the clog.

THAWING FROZEN PIPES

You may think that all of your plumbing system is in perfect working order and that there is little or no chance of a pipe bursting and flooding your house. Yet there is one situation that you may not have considered. Water that freezes during

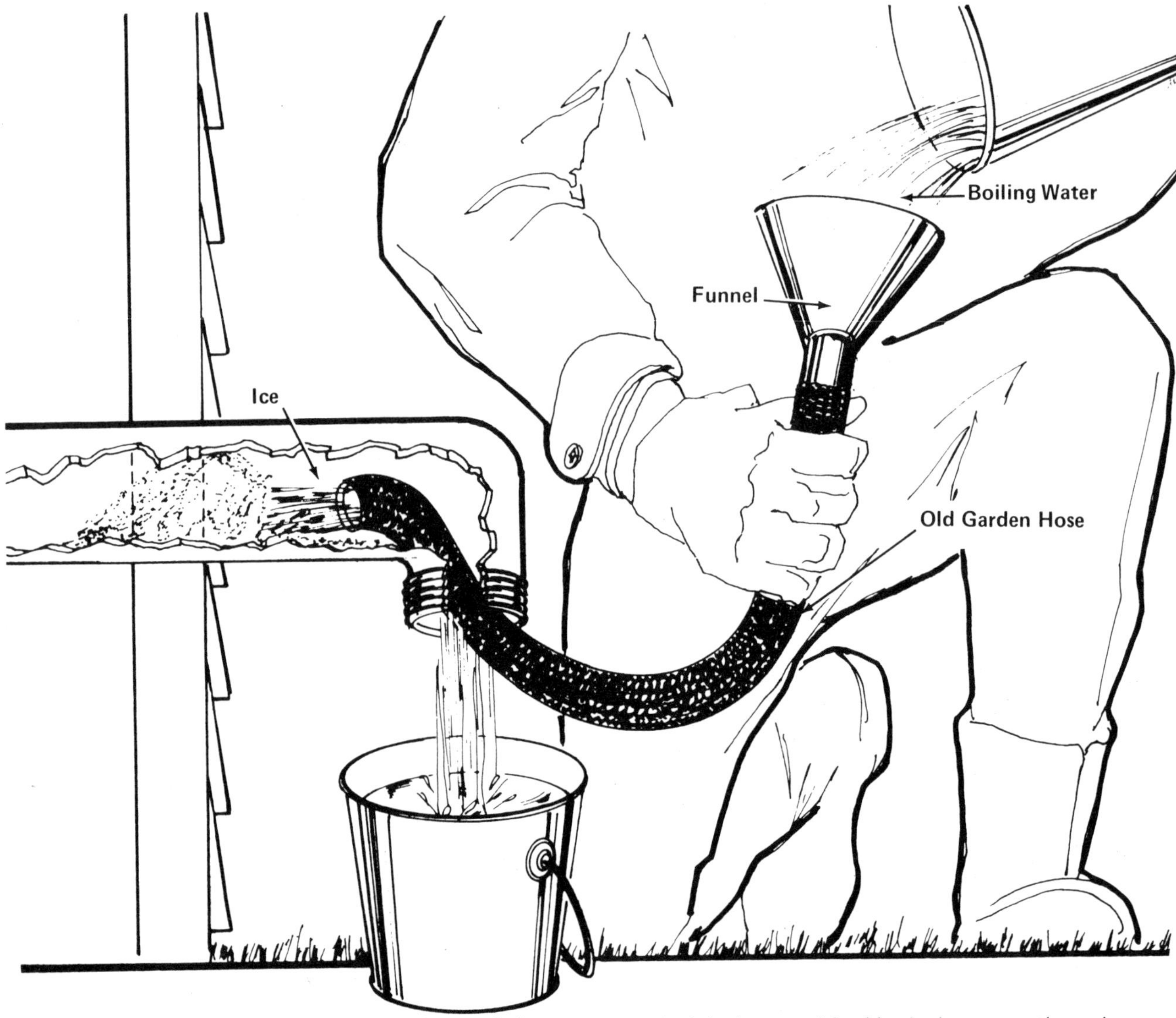

There is a special technique for clearing a frozen drain pipe which cannot be thawed merely by pouring hot water down at the drain opening. What you need to do is get the hot water closer to the ice blockage before the water has a chance to cool. Remove the trap and insert a short length of garden hose into the pipe. When you cannot push the hose in any further, you know that you have reached the problem area. Raise the outside end of the hose and feed in the hot water through a funnel. That is the best way to get the hot water directly to the blockage. But be careful! Until the ice melts and goes down the pipe, the hot water that you pour through the hose can back up, splash, and scald you. Pour the water in gradually, and have a bucket ready to catch the back-flow.

winter in an unprotected pipe expands, and that expansion can rupture an otherwise sound pipe. Of course, a frozen pipe is always an inconvenience, but it can result in a much more serious situation than just a temporary loss of water.

If ever there were an example of how a little prevention is far superior to a lot of cure, then fighting frozen pipes is it. By taking the proper preventative steps, you may never need to worry about thawing frozen pipes — or worse yet — repairing a pipe that ruptures when the water in it freezes solid.

If you are building a new house or adding pipes underground, follow these freeze-fighting tips. Bury the pipes below the frost line if possible. Since this depth is different in different parts of the country, check with your local U.S. Weather Bureau office to learn about the frost line in your community. Make sure that the new pipes are as well-insulated as is practical, and run as many pipes as you can along your home's inner walls instead of along its outside walls. If you have pipes on outside walls leading to outside faucets, make sure that the faucets are the frost-free type. When the cold weather sets in, cut off the outside faucets and open the tap to drain the exposed pipe and faucet. Finally, if you know that you have a stretch of pipe that is subject to freezing, wrap the pipe with heat tape (sometimes called heat cable). You can purchase heat tape that has an automatic thermostat to start the heat when the temperature outside drops to about 35 degrees.

What happens, though, when you wake up some frigid winter morning to find a water pipe frozen solid? Then the suggestions for preventing such an unfortunate occurrence will be of little value or consolation. What you will want then are not freeze-fighting tips, but rather directions on the best way of thawing frozen pipes.

Before doing anything else, open the tap so that the steam produced by your thawing activities will be able to escape. In addition, start thawing the pipe at the tap and work back toward the origin of the freeze. As you melt the ice, the water and/or steam will come out the open tap; if you were to start in the middle, the steam produced by the melting ice could get trapped and build up enough pressure to rupture the pipe.

You can use any of several pipe-thawing heat sources. Probably the most popular — and certainly the safest — is simply hot water. Try wrapping a heavy towel or a burlap bag around the pipe to concentrate and hold the heat against the pipe. Pour the hottest water you can get over the towel, but be careful because most of the water will run off of the towel onto you or the floor. A properly positioned bucket can save you both a scalding and a mess.

A less messy but far more dangerous heat source for thawing frozen pipes is a propane torch. You must be extremely careful to prevent the torch's flame from igniting the wall behind the pipe. A scrap of asbestos siding or some other fireproof material between the pipe and the wall is a good precautionary measure, but the way you use the torch is the main ingredient in safe pipe thawing. Keep the flame moving back and forth; never leave it very long in one spot. Be especially careful when you come to any soldered joints. Pass over them very quickly or you will find that you have a much more serious plumbing problem on your hands than just a frozen pipe.

If you want to avoid the messiness of thawing with hot water and the danger of melting with a propane torch, you can use a heat lamp or an electric iron as the heat source. Both are very slow, but both are also very safe.

There is a special technique for thawing a frozen drain pipe. Actually, all it involves is melting the blockage with hot water; but if the ice is at some distance from the drain opening, the hot water you pour in could cool considerably before it reaches the trouble spot. Therefore, in such cases you should remove the trap and insert a short length of garden hose into the drain pipe. When the hose reaches the ice — that is, when you cannot push it along any further — raise your end of the hose and feed hot water in with the aid of a funnel. This way, the hot water is sure to get to the problem area; but you must be careful when using this technique. Until the ice melts and drains down the pipe, the hot water you pour in will back up at you. Have a bucket ready under the pipe to catch the overflow, and do whatever is necessary not to scald yourself.

QUIETING PIPES THAT BANG

Most people refer to what they hear when pipes bang as "water hammer," but water hammer is only one of several different mysterious noises that can come forth from the plumbing in your home. If you hear the sound whenever you turn on the water, then the problem is certainly not water hammer; the banging probably results from the pipes hitting against something.

Banging pipes are much easier to cure if you can see them. Therefore, turn on the water and start looking for movement; once you find the trouble, you should have no problem in stopping the pipe or pipes from hitting against whatever is in the vicinity. Yet, even if the moving pipe is between the walls, you can still silence it without tearing your house apart. Just place stops at each end where the pipe emerges from behind the wall.

Generally, you will find that the moving pipe is loose within its tube strap or U-clamp and is banging against the wall to which it is supposed to be secured. To eliminate the noise completely, just slit a piece of old garden hose or cut a patch of rubber and insert it in the strap or clamp to fill in the gap. Pipes that strike against a masonry wall can be silenced by inserting a block of wood between the pipe and the wall. Nail the block to the wall with masonry nails — or with a screw installed in a masonry wall anchor — and then attach the pipe to the block with a pipe strap.

In a basement or crawl space, you will frequently find pipes suspended from the joists with perforated pipe straps. While this is a perfectly proper installation, a long run of suspended pipe can move within the straps, strike against something, and create a racket. A block of wood placed strategically along the run will eliminate the pipe's movement and, of course, the noise. Whenever you secure a pipe, however, you must be careful not to anchor it so as to restrict its ability to expand and contract with the changes in temperature. If you place a bracket on a pipe, be sure to include a rubber buffer between the pipe and the clip. You can make such buffers from garden hose, radiator hose, foam rubber, rubber cut from old inner tubes, or even kitchen sponges.

You may find that supply pipes and drain pipes that run right next to each other are striking one another and creating a clatter. Your best solution to this problem is to solder the two pipes together;

that should silence the knocking you hear. If the knocking sound occurs only when you turn on the hot water, it means that the water heater is set too high; the noise is the steam that is rumbling throughout your hot water system. Just turn the heat setting down and you will silence these particular pipes.

Pipe that is either too small to begin with or that has become restricted with scale or mineral deposits can be a big noise problem. It is almost impossible to clean clogged supply pipes (a water filter can sometimes prevent such clogging), and you must replace pipe that is too small if you want to stop the noise. On the other hand, you can diminish the sound level of clogged pipes considerably by wrapping them with sound-damping insulation.

Drain pipes rarely clatter, but they can emit a sucking sound as the water leaves the sink or basin. Such a sound means that a vent is clogged or, perhaps, that there is no vent at all attached to the drain. In either case, you have a potentially serious plumbing problem on your hands — not because of the noise, but because a nonfunctioning or nonexistent vent could result in the water seal getting siphoned from the trap, allowing sewer gas to back up into your home. Run a plumber's snake (or ribbon) through the vent to eliminate any clogging; but if there is no vent on the drain, install an anti-siphon trap (available at your hardware store) to quiet the noise and to prevent any problem with sewer gas.

Water hammer is a particular type of plumbing noise, not a generic name for any kind of pipe clatter. It occurs when you shut off the water

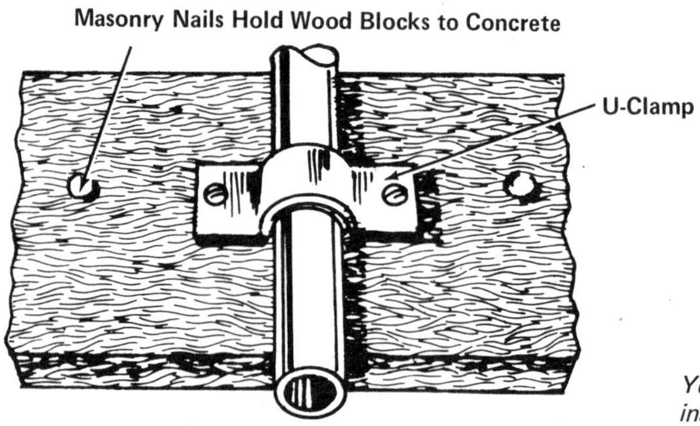

There are several ways to stop pipes from striking walls. If the wall is masonry, put a wood block behind the pipe, and fasten it with masonry nails.

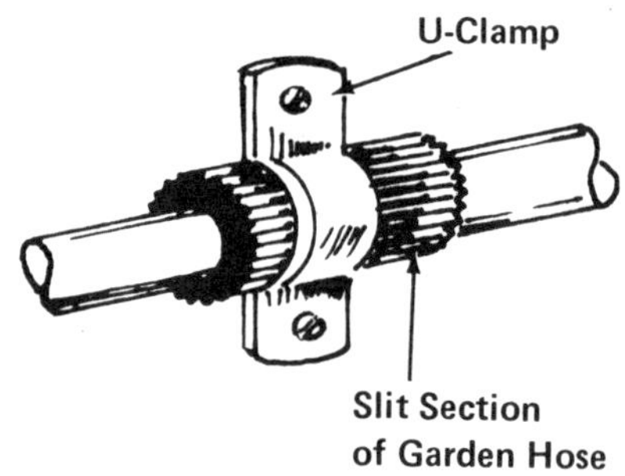

You can stop a pipe from moving within a U-clamp by inserting a piece of old garden hose in the clamp.

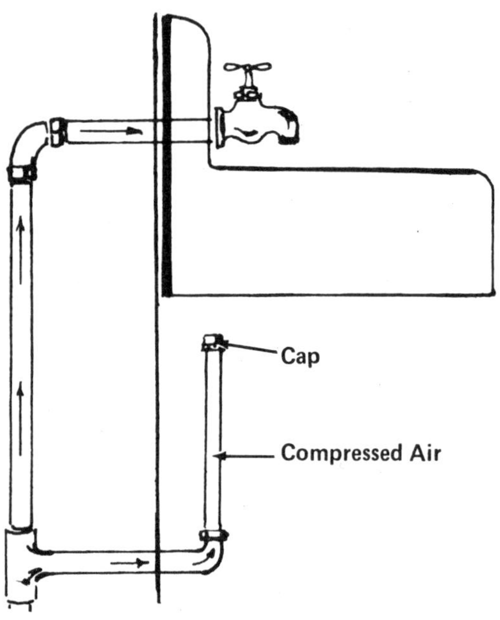

Properly installed plumbing possesses air chambers to absorb the shock when the water flow is cut off.

If your plumbing lacks built-in air chambers, you can add substitute devices without tearing into walls.

Pipe Troubleshooting Chart

PROBLEMS	CAUSES	REPAIRS
Leaking Pipe	1. Joints not watertight.	1. Tighten if possible. Remove and cover threaded joints with pipe dope. Remove and resolder sweat joints.
	2. Hole in pipe.	2. Place clamp-type repair kit over hole. Replace bad section of pipe or reroute new tubing if bad pipe inaccessible.
	3. Burst pipe.	3. Shut off water at main cutoff. Repair leak. Avoid electrical shocks due to contact between appliances and water.
Pipe drips water, but no leak	1. Condensation.	1. Wrap pipes with no-drip material.
Noise in pipes — hot water only	1. Steam causes rumbling in hot water pipes.	1. Turn down thermostat setting on hot water heater or replace faulty thermostat.
Water has sucking noise when draining	1. Improper venting.	1. Clean vent on roof. Add antisiphon trap if no vent is there.
Hammering noise when water is shut off	1. Air chambers water logged.	1. Shut off and drain supply line to allow air to reenter.
	2. No air chamber.	2. Install air chambers.
Banging noise while water running	1. Loose pipe.	1. Track down which pipe is loose and brace, cushion, or strap it.
No water supply	1. Frozen pipes.	1. Open taps. Start thawing at closest point to tap and work back.

suddenly, and the force of the fast-moving water rushing through the pipe is brought to a quick halt — creating a sort of shock wave and a hammering noise. Plumbing that is properly installed possesses air chambers which compress when the shock wave hits, softening the blow and preventing any hammering. The chambers can fail, though, because water under pressure gradually absorbs the air. If you had formerly experienced no hammering and then it suddenly started, most likely your plumbing system's air chambers became waterlogged.

You can cure water hammer by cutting off the water beyond the waterlogged chambers (perhaps even at the main cutoff), opening the offending faucet, and permitting the faucet to drain thoroughly. Once all the water drains from the chamber, air will go back in, fill up the chamber, and restore the cushion. If the air chamber is located below the outlet, you may have to drain the main supply lines to allow the chamber to fill with air again.

The air chamber will not drain properly if it is clogged with scale or residue from chemicals in the water. The chamber should be a tiny bit larger than the supply pipe to which it is attached in order to preclude such clogging. Since the chamber is merely a capped length of pipe, however, all you have to do to clear it is remove the cap and ream out the scale.

What do you do if there are no air chambers built into your plumbing system? The answer is that you either must learn to live with the hammering or else install the necessary air chambers. If you have no room to make the installation (without tearing into a wall), go to your plumbing supply dealer and find out about the substitute devices that are designed for such problem areas. Many of these devices have a valve that makes it easy for air to re-enter the system.

If you are adding new plumbing, make sure that the pipes are big enough to allow the water to flow properly. Many times, the minimum size acceptable to your local plumbing code turns out to be unacceptable to you when you hear it clatter. Remember to keep the air chambers in good working order and to run pipes in such a way as to prevent their striking each other or nearby walls. If you follow these simple procedures, you should be able to cure most plumbing noises before you ever hear them.

Chapter Three

Toilets

The toilet may not be the most pleasant subject to discuss, but it is certainly one of the most important fixtures in your home. Although toilets are — if not maltreated in any way — sturdy and reliable components of the plumbing system, it is a rare homeowner or apartment-dweller who never has any problems with the john. Clogging is, perhaps, the most serious toilet trouble, but it is far from the only one. The tank, for example, can make all sorts of weird noises, or it can run water continuously. The commode can crack, or you may spot leaks around the bottom where the fixture goes into the floor. Fortunately, most toilet troubles can be fixed by any well-informed do-it-yourselfer.

REPLACING A TOILET SEAT

Undoubtedly, the easiest toilet task is replacing the seat. There are so many styles of replacement seats available that you should encounter no difficulty in finding one to match any bathroom color scheme or motif. Most modern toilets are manufactured to a standard size, and replacement seats are made to fit that standard. If your toilet is extra wide or very old, however, you may not be able to use a standard size replacement seat. In such instances, you may have to place a special order for the seat with a company that deals in plumbing fixtures.

Once you have the right size seat, you are faced with the problem of removing the old one. Removal sounds simple. All you must do is remove two nuts and lift your old toilet seat up and out. You may ask, "What is so difficult in doing that?" Wait until you try to loosen those two nuts. Consider yourself among the luckiest of individuals if the nuts that hold your toilet seat are not rusted or corroded; sometimes, in fact, they are recessed and practically inaccessible.

If you can get to them relatively easily, apply some penetrating oil to help loosen the nuts. Give the oil plenty of time, and then be sure that you do not use too much muscle if they are still stubborn. Were the wrench to slip off the stubborn nut, it could crack the tank or the bowl or anything else it happened to strike. If the nuts seem to be out of your reach with a regular wrench, try a deep-throated socket wrench.

When all else fails, reach for your hacksaw, insert it under the hinge, and saw right through the

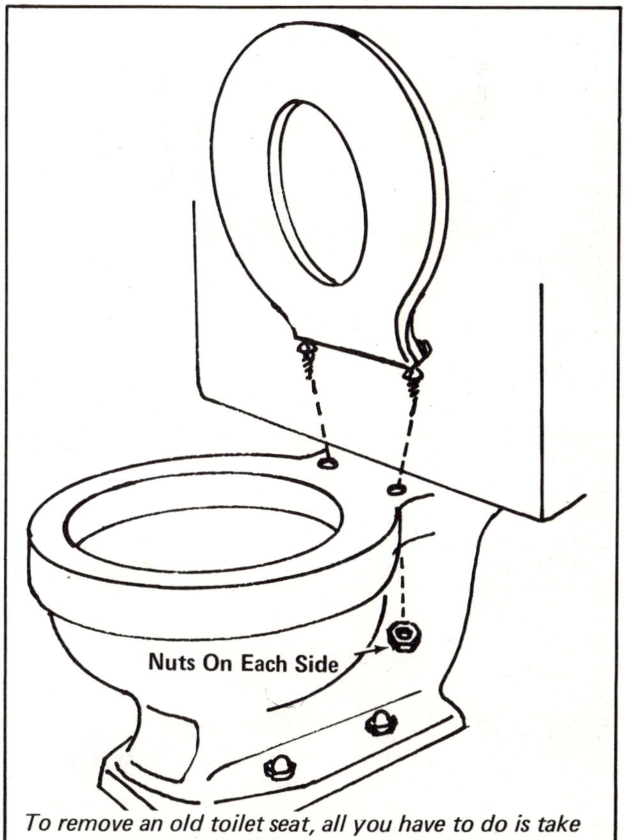

To remove an old toilet seat, all you have to do is take out two nuts and then lift the seat up and out.

bolts. Sawing is, however, a measure of last resort; try every other procedure to loosen the nuts first before you pick up the hacksaw. Then, if you are convinced that nothing is going to break the nuts loose, be extremely cautious in using the saw. A careless slip with a hacksaw can crack the fixture just as a wrench that comes flying off a stubborn nut can crack it.

With the nuts removed or the bolts cut, you can remove the old seat without further difficulty. The new one is fastened simply by replacing the bolts and tightening the nuts, but be careful not to overtighten the nuts — you may want to replace this seat too some day. If you live in an apartment and put on one of the new mod seat and lid units, be sure to keep the old one. When you are ready to leave, put the old toilet seat back on and take your fancy new one along with you.

If the lid and seat are still in good condition, but the little rubber bumpers on the bottom are shot, you can buy replacements at your local hardware store. Some of these bumpers screw in, while others require that you nail them into the underside of the seat. Whichever type you have, try to install the new ones in holes that are close enough to the old ones that the new bumpers cover the old holes.

Toilet Tools

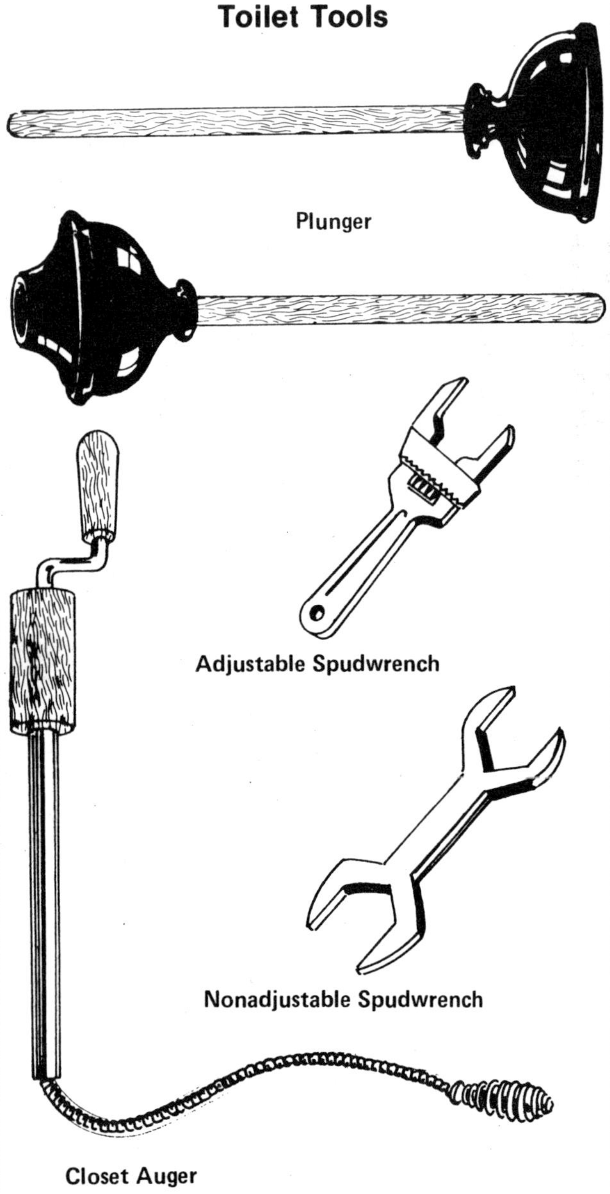

Plunger

Adjustable Spudwrench

Nonadjustable Spudwrench

Closet Auger

TOILET TOOLS

Few special tools are required for changing the seat: a wrench, or perhaps a deep-throated socket wrench or hacksaw. Most do-it-yourselfers have a wrench of some sort and a hacksaw, but you may not own a socket wrench set. If you do not already have a set, it does not pay to buy one just for this job; but you will find many occasions to use a socket wrench set in addition to your do-it-yourself plumbing chores.

Of course, the most frequently used toilet tool is the plumber's friend, also referred to as a plunger or force cup. Get one with a long handle, and be sure that the suction cup is large enough to cover the drain opening.

A closet auger is a version of the plumber's snake, but it is designed specifically for clearing clogs in toilets. The closet auger is shorter than a regular snake, and it comes encased in a metal housing with a crank.

If you really get involved in plumbing chores and you decide to remove a toilet either for replacement or repairs, you may well need a tool called a spudwrench. Older toilets frequently have a large pipe that connects the tank to the bowl; the pipe is called a spud, and it is held to the bowl and tank by slip nuts. A spudwrench is designed to remove these extra-large hexagonal slip nuts. The adjustable type is far more versatile than the nonadjustable spudwrench, which has wrenches of different sizes at either end.

CLEARING CLOGGED TOILETS

You can generally clear a clogged toilet with the plumber's friend, or plunger. Just make sure that there is enough water in the bowl to cover the rubber suction cup, and then work the handle of the plunger up and down. If there is not enough water in the bowl, do not flush the toilet; flushing a clogged toilet will just cause the bowl to overflow. Instead, haul in a pan of water from another source to supply the water you need to cover the plunger cup.

Usually, whatever is blocking the toilet drain is not very far away. Therefore, when you decide that you have had enough arm exercise pumping the plunger up and down without success, you can go on

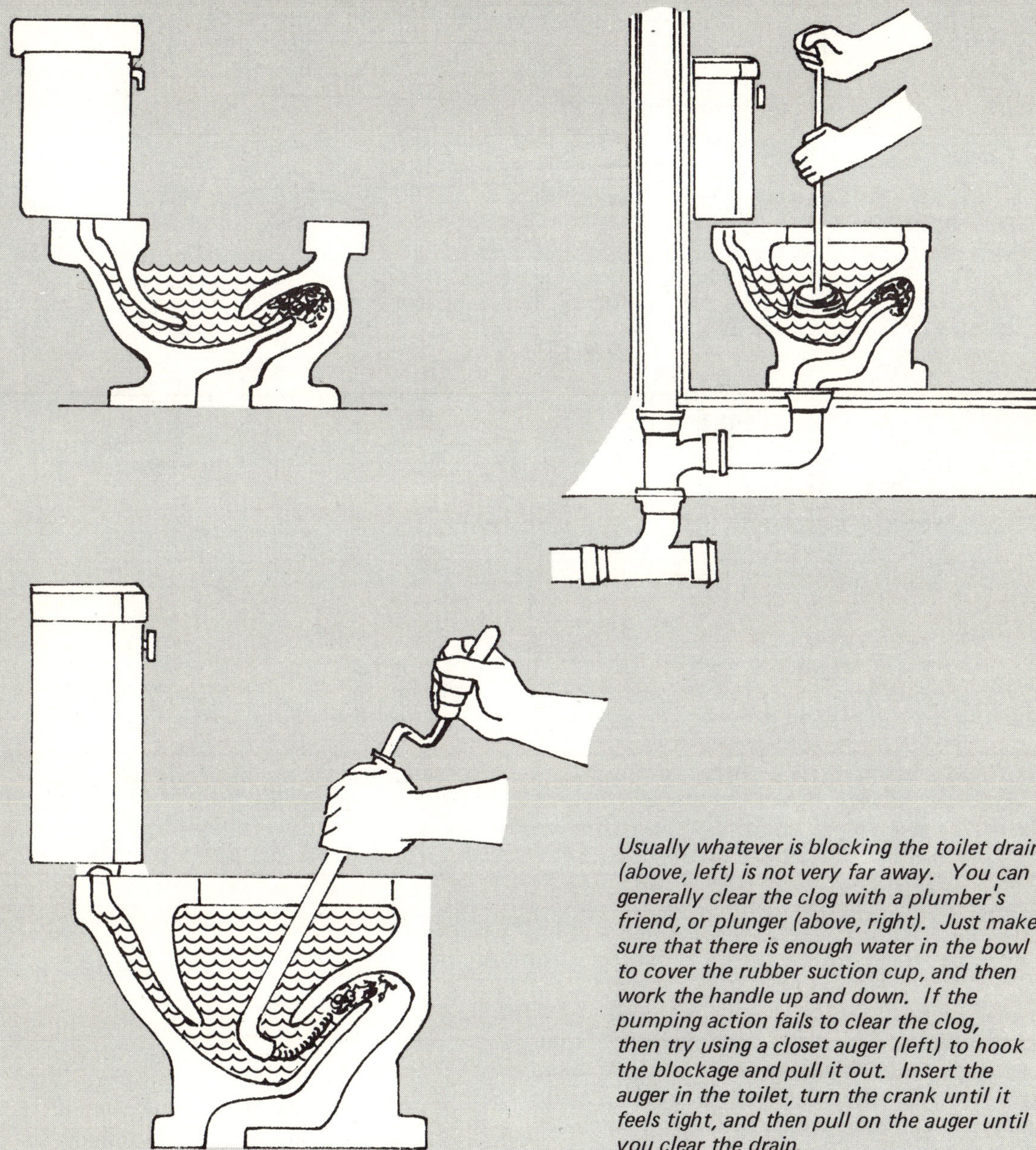

Usually whatever is blocking the toilet drain (above, left) is not very far away. You can generally clear the clog with a plumber's friend, or plunger (above, right). Just make sure that there is enough water in the bowl to cover the rubber suction cup, and then work the handle up and down. If the pumping action fails to clear the clog, then try using a closet auger (left) to hook the blockage and pull it out. Insert the auger in the toilet, turn the crank until it feels tight, and then pull on the auger until you clear the drain.

to the next step: trying to hook the blockage and pulling it out from where it is lodged. A wire coat hanger often can do the job adequately, but the coat hanger is really a substitute for the plumbing tool expressly designed for these situations: the closet auger.

Insert the auger in the toilet and turn the crank until it feels tight. This means that the snake has twisted its way to and into the blockage. Pull on the auger and you should be able to remove whatever is clogging the toilet. Try the closet auger several times if you are not successful at first, but you may have to resort to pushing a regular plumber's snake through the blockage if you are unsuccessful in pulling it out with the auger.

When all else fails, you may have to remove the toilet from the floor and turn it upside down to get a straight shot at the blockage. Naturally, you should give all the other methods as much opportunity as possible before you actually remove the toilet; it is not what anyone would call a fun task. It is not, however, beyond the capabilities of the average do-it-yourselfer, and we will explain exactly how to do the job in a later section.

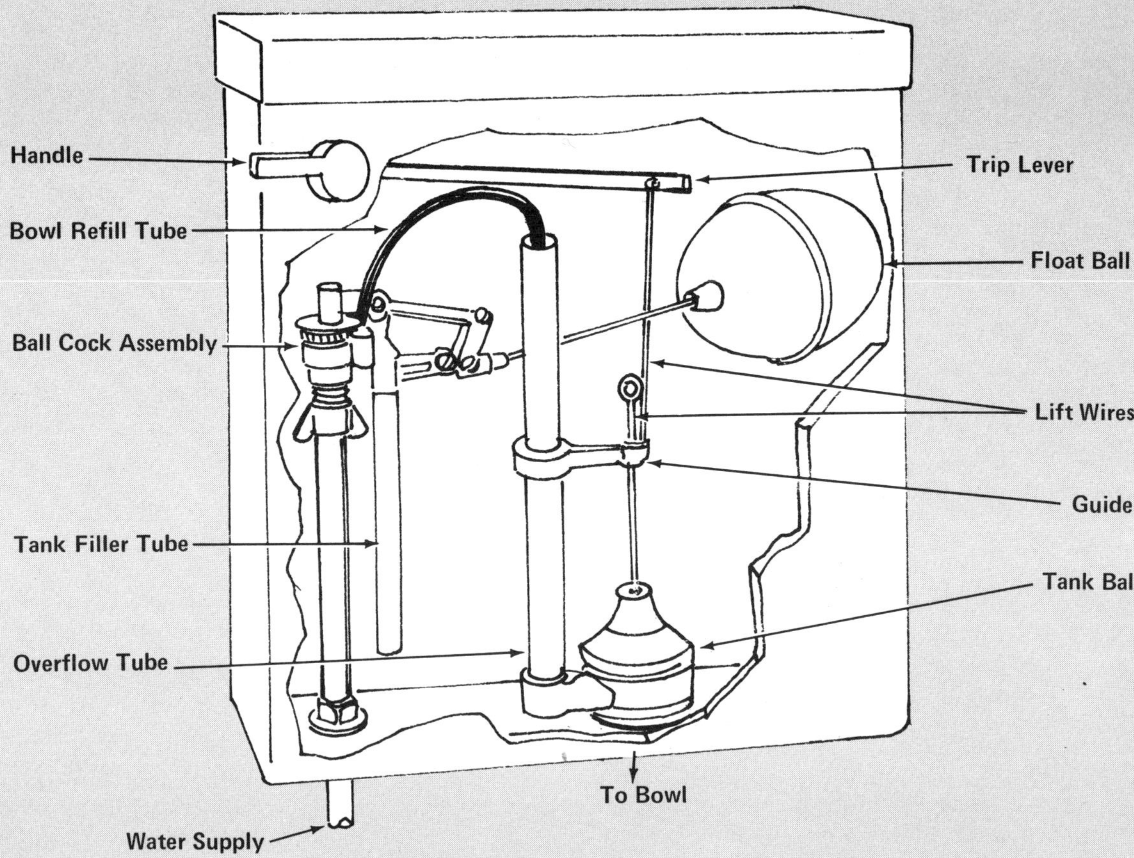

The inner workings of a toilet tank are really quite simple and logical, and you can repair most tank troubles easily.

TOILET TANK TROUBLES

Compared with a clogged toilet, tank troubles can seem relatively insignificant. Yet these troubles — such as weird noises or water running continuously — can be more than annoying; they can also be costing you a good sum of money in wasted water. Fortunately, you can eliminate most toilet tank troubles quickly and easily.

Ever wonder exactly how a toilet works — what happens between the time that you operate the handle and the time that the toilet is fully restored for further use? Before you can solve the problems that plague your toilet, you must know exactly how this most vital of fixtures operates. Follow these steps to understand what happens when you flush a toilet:

1. You trip the handle on the outside of the tank.
2. The handle raises the trip lever inside the tank.
3. The lift wires, connected to the trip lever, are lifted as the trip lever rises.
4. The tank ball, fastened to the bottom lift wire, rises.
5. The water in the tank rushes out past the raised tank ball and into the bowl below.
6. The rush of water raises the bowl's level above that of the water in the drain spout. Since water seeks its own level, the water from the tank pushes the bowl water out into the drain and causes a siphoning action that cleans everything out of the bowl. When all the water is gone from the bowl and air is drawn into the drain spout, the siphoning stops.
7. Meanwhile, the tank ball is no longer supported by the water in the tank. Consequently, the ball falls back in place, covering the flush valve opening.
8. In addition, while the water was rushing out of the tank, the float ball (which floats on top of the water) dropped down. The dropping float ball pulled down on the float arm, raising the valve plunger and permitting new water to flow into the tank.

9. As the water level rises in the tank, the float ball rises until the float arm gets high enough to lower the valve plunger and shut off the water. If the water should fail to shut off for some reason, there is an overflow tube that carries the water down into the bowl to prevent the tank from overflowing.

Lift the lid off your toilet tank, and you should be able to follow this procedure quite easily. Once you know how the toilet works, you can start to search for the source of your toilet tank troubles.

Suppose that your toilet never stops running — water flows continuously out of the tank to the bowl and out the drain. What should you do? Lift up on the float arm. If the water stops, then you know that the problem is that the float ball does not rise far enough to lower the valve plunger. One reason that the float arm does not rise far enough could be that the float ball is rubbing against the side of the tank. If it is rubbing, just bend the float arm slightly to move the ball away from the side.

The problem could be in the ball itself. If the ball does not touch the tank, continue to hold the float arm and remove the ball by turning it counter-clockwise. Shake the ball to see if there is water inside; the water could be preventing the ball from rising normally. If there is water in the float ball, replace the ball with a new one. If there is no water, put the old ball back on and bend the float rod — very gently — to lower the level which the float ball must reach to cut off water entry into the tank.

What do you do if, when you lift the float arm, the water does not stop running? You know then, of course, that the problem is not in the float arm at all. The next logical place to check for trouble is the tank ball. Frequently, chemical residue from the water can prevent the ball from seating properly, or the ball itself may have decayed. The result is the same in either case — water will seep around the flush valve opening into the toilet bowl below.

Shut off the water at the toilet cutoff, and then flush the toilet to empty the tank. You can now examine the tank ball for signs of wear, and install a new ball if so indicated. If the problem is chemical residue on the lip of the flush valve opening, take some wet-dry emery paper, steel wool, or even a knife and clean away the debris. You can retard any further formation of such residue by depositing leftover soap slivers in the toilet tank.

There are still other possible causes for a toilet running continuously. The guide or the lift wires that raise and lower the tank ball may be out of line or bent. Make sure that the lift arm is not rubbing against anything and that the lift wire is not installed in the wrong hole of the arm; either situation could cause the tank ball to fall at an angle and not block the opening as it should. If neither the float ball nor the tank ball is at fault, then the problem must be in the ball cock assembly.

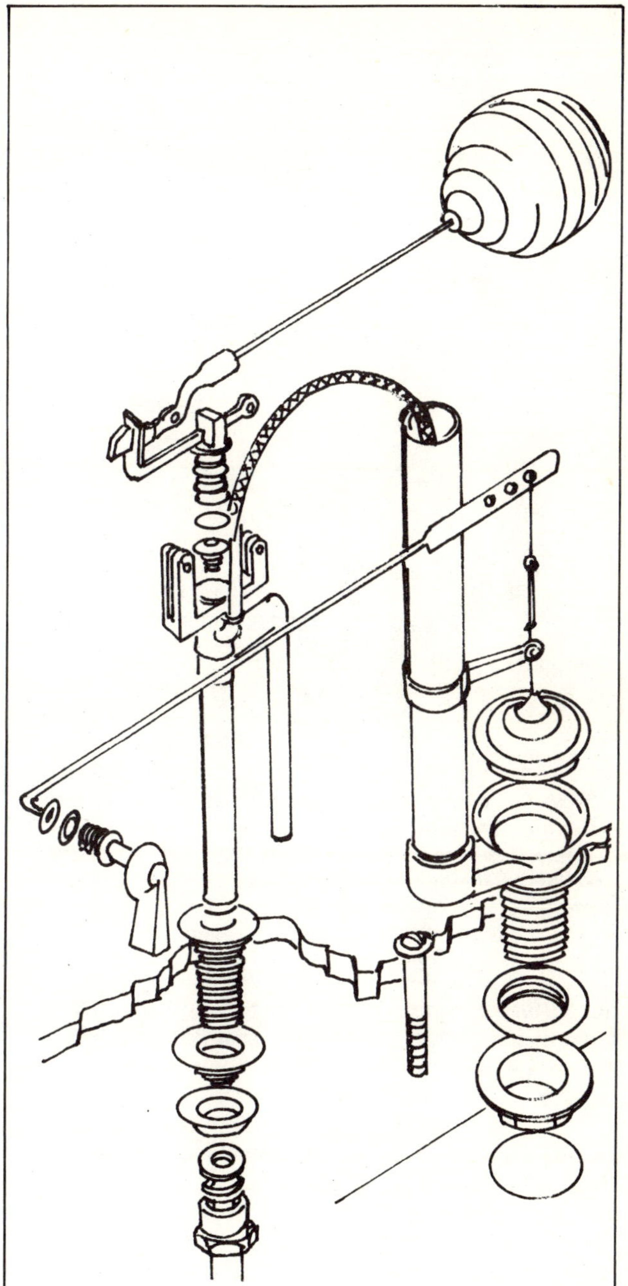

When you trip the handle, the trip lever inside the tank lifts the tank ball. Water rushes out, lowering the float ball and float arm which in turn raise the valve plunger, permitting fresh water to enter the tank.

WORKING WITH THE BALL COCK ASSEMBLY

The ball cock assembly looks far more complicated than it really is. Just make sure that the water cutoff valve for the toilet is in the off position, and you are ready to go to work. With most ball cock assemblies you will find a pair of thumb screws that

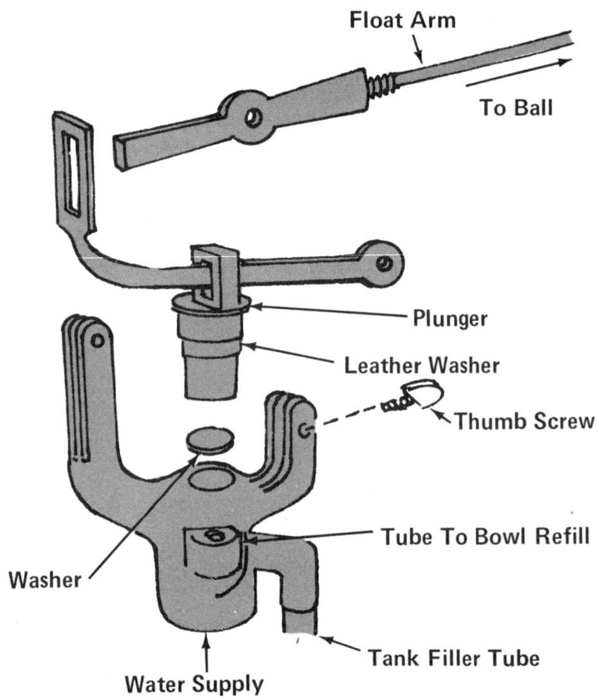

The ball cock assembly can be responsible for many of the toilet tank's troubles. If the toilet runs water continuously, the problem is probably due to a faulty washer on the ball cock's valve assembly.

hold the valve plunger; but if the unit in your toilet tank has a different linkage arrangement, you still should be able to figure out how to remove the valve plunger.

Remove the valve plunger, and you will see one or two washers — sometimes you will find an "O" ring. Naturally, if any of these are faulty, water will flow out past the plunger continuously and the toilet will run without ceasing. Examine all of the washers, and replace any defective ones.

Some ball cock assemblies are completely sealed, however, and there is no way that you can get inside the unit without breaking it. If that is what you find when you lift the lid off the toilet tank, then you must buy a new ball cock assembly. Most replacement units cost only a few dollars.

Should you discover either a sealed ball cock assembly or a damaged one, the first thing to do is to get all the water out of the tank so that you can remove the old unit and install a new one. Shut off the toilet's water supply at the cutoff valve, flush the tank, and then use a large sponge to remove whatever water remains in the tank. Unscrew the float arm from the old ball cock unit, and remove the refill tube from the overflow tube (the refill tube may be clipped on or it may just be bent into position).

Now, look under the tank. You will see a coupling nut where the water pipe meets the tank. Loosen both the coupling nut and the lock nut that is right above it. At this point, you can lift out the old ball cock assembly from the tank, but be sure to save all the washers from all connections both inside and outside the tank. You may not need them — most likely all the necessary washers are included with the new unit — but it is always smart to hang onto the old parts until you install the new ball cock assembly.

Insert the new unit in the hole in the tank with the inside washer in place. Tighten the lock nut on the outside sufficiently to make the inside washer fit watertight against the hole, but do not overtighten it. Replace the coupling nut, reinstall the float arm, and the job is done. Turn the water back on at the toilet cutoff valve, and check for leaks at all joints. Of course, another thing to check for is that the float ball does not rub against the back of the tank.

When you go to your hardware or plumbing supply store in order to purchase a new ball cock assembly, you will find that you must decide between plastic and metal units. Plastic costs less and will not corrode, but plastic assemblies are not as sturdy as metal ones. In addition, plastic units usually cannot be repaired because most of them are sealed. Nevertheless, you can purchase a different type of unit than the one you are replacing so long as the new assembly has a threaded shank the same size as the old one.

The newest types of ball cock assemblies eliminate the float arm and the float ball. Fluidmaster — the most popular model — has a plastic cup that floats up to cut off the water as the tank fills. You can set the water level in the tank by adjusting the position of the plastic cup on the rod. One big advantage of this type of ball cock assembly is that it lets the water run full force until the tank is filled; then it shuts the water off immediately, eliminating any groaning noises that some toilets emit as their conventional float arms gradually close the valve.

Another new replacement for the ball cock assembly is a unit called the Fillmaster Fill Valve, which — like the Fluidmaster — eliminates the float ball and float arm. A very small unit, the Fillmaster rests almost on the bottom of the tank, and its diaphragm-powered valve senses the level of the water from down there. Moreover, since it requires no tools, it is an easy unit to install. You may need a couple of common tools for the removal of the old ball cock assembly, but after that all you do is follow these steps:

1. Cut off the water.
2. Flush, and then sponge out the remaining water from the tank.
3. Remove the old ball cock assembly.
4. Slip the following parts over the supply pipe under the tank: coupling nut, friction washer, cone washer, and mounting nut. Put these parts — all included with the Fillmaster Fill Valve — on in order.

5. Put the unit in place in the tank, fitting the threaded shank over the supply pipe. Make sure that the gasket shank fits down into the hole.
6. Now you are ready to put it all together. Do not use any pipe joint compound. Start the mounting nut on the threaded shank section, and hand-tighten only.
7. Push the washers into place, and hand-tighten the coupling nut.
8. Attach one end of the refill tube to the overflow pipe, and place the other end on the stem of the Fillmaster unit. You are now ready to turn the water back on, and the tank will start filling. The Fillmaster is preset to a level of about eight inches, but you can adjust the level with the knob marked "ADJ."

The plastic Fillmaster Fill Valve unit carries a lifetime guarantee, but the guarantee requires that you mail the faulty unit back to the factory. Unless you save your old ball cock assembly, your toilet will be out of commission until the mail delivers the new unit.

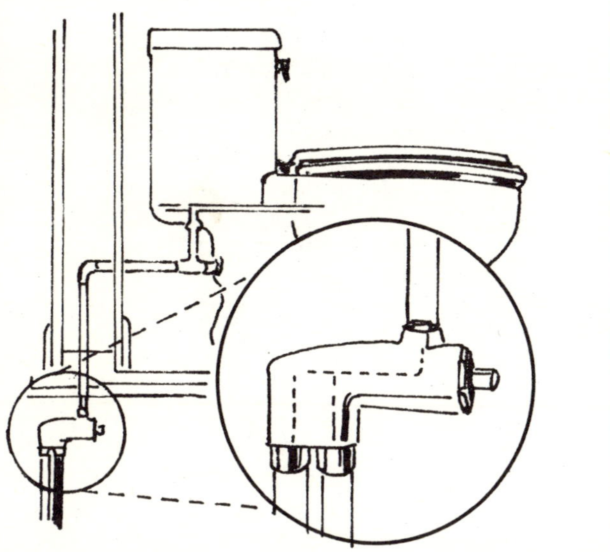

One of the best ways to eliminate toilet tank sweating (condensation) is to install a temporator valve. Such a valve provides a regulated mixture of hot and cold water, thereby minimizing the temperature difference between the water inside the tank and the surrounding air outside.

INADEQUATE FLUSHING, SWEATING, AND OTHER PROBLEMS

What can you do if too little water comes from the tank to flush the bowl clean? The first thing to check is the water level in the tank; it is probably too low. If the water level fails to reach within 1½ inches from the top of the overflow tube, then bend the float arm up a little to allow more water to enter the tank.

In some cases, the water level may be correct, but there still is an insufficient amount of water coming from the tank to clean the bowl properly. Most likely, the culprit in this situation is the tank ball. The ball is probably dropping too soon, and the reason it is doing so is that the guide is set too low. Raise the guide, but make sure that it stays in line with the lift wires. If the guide and the wires are out of alignment, then the tank ball will not drop straight into the hole, and you will find that you have a toilet that runs continuously.

There is one other possible cause for inadequate flushing. The small ports around the underside of the bowl's rim can get clogged with residue from chemicals in the water and prevent a sufficient amount of tank water from running out into the bowl. A small purse mirror can help greatly in seeing the holes, and a piece of wire coat hanger or an offset Phillips screwdriver — if you have one — can ream out any clogging that may have occurred.

Toilet tanks can sweat and drip onto your floors just as pipes can. There are jackets that are designed specifically to fit over the tank and to absorb the moisture, and there are drip pans that fit under the tank to catch and collect the drops so that they do not damage your bathroom floor. A new gadget, called a temporator valve, is perhaps the best way to eliminate tank sweating. The temporator valve provides a regulated mixture of hot and cold water, which lessens the difference between the temperature inside the tank and the temperature of the surrounding air. It is that temperature difference, of course, that causes condensation — or sweating.

A temporator valve requires that you hook up a hot water line to the valve, which may be quite inconvenient if there is no such line relatively close to the toilet. Moreover, the temporator valve does not prevent the water inside the tank from cooling between flushings; thus, condensation can still occur even on a temporator-equipped toilet. Fortunately, there is yet another method for stopping a toilet tank from sweating. Although it works nearly every time, some people may consider this method a little strange. It involves buying one of the small heater units designed for a fish tank and rigging the heater to fit the tank of your toilet. Since it heats the water continuously, you should never again be troubled by a sweating toilet tank.

One of the worst effects of a sweating tank is that it can hide a genuine leak in the tank. You may see water on the floor, but you fail to attribute it to a leak, believing that it is merely some condensation that dripped off. There is, fortunately, an easy way to check for leaks. Pour enough bluing into the tank to give the water a noticeably blue color; some of the toilet bowl cleaners that fit inside the tank also turn the water blue. Then, if the tank does have a leak, the moisture on the floor will show traces of blue and you will know that it is time to replace the tank with a new one.

Sometimes do-it-yourselfers create more problems than they cure when they tackle toilet repairs. If you follow a few simple guidelines, you could save yourself plenty of grief — and money — when you work on the family john. First, when you remove the tank top to see what the problem is, put the top on the floor at a safe distance from where you will be working. The idea is to minimize the chance that you could knock the tank top off from any height (say, the bowl or the sink) onto the floor or that you could inadvertently step on it. If you were to do either one, you would probably have to go out and buy a new tank top — a fine way to lose all the money you saved by taking care of the toilet tank trouble yourself.

Second, whenever you encounter a stubborn nut, use penetrating oil on the nut and do not strain to loosen it with a wrench. If you strain, there is a good chance for the wrench to slip off the nut and crack the tank or the bowl. Similarly, avoid overtightening tank nuts; too much pressure can also crack the toilet tank. Once you crack the toilet, then you must study the next section carefully.

REMOVING AND REPLACING A TOILET

Removing and replacing a toilet is not a task to be undertaken without good reason, but it is certainly not beyond the capabilities of the well-informed do-it-yourselfer. Perhaps you have grown tired of the old john's appearance; or you may have

Toilet Troubleshooting Chart

PROBLEMS	CAUSES	REPAIRS
Water in tank runs constantly	1. Float ball or rod catching.	1. Look to see if ball is touching back tank wall. Bend float rod gently to move ball or rod.
	2. Float ball not rising high enough to shut off valve.	2. Gently bend rod down, but only a little.
	3. Tank ball not seating.	3. Check lip of valve, and scrape away any corrosion. Replace tank ball if worn. Adjust lift wires and guide if ball cannot fall straight.
	4. Ball cock valve does not shut off.	4. Washers need replacing, or entire unit needs replacing.
Nonflush or inadequate flush	1. Clogged drain.	1. Remove blockage.
	2. Not enough water in tank.	2. Raise level of water by gently bending float rod up.
	3. Tank ball falls back before enough water leaves tank.	3. Move guide up so that tank ball can rise higher.
	4. Leak where tank joins bowl.	4. Tighten spud nuts or replace spud washer.
	5. Condensation.	5. Install tank cover, drip catcher, or unit to bring hot water to tank.
Tank whines when filling	1. Ball cock valve not operating properly.	1. Replace washers or install new ball cock unit.
	2. Water supply restricted.	2. Be sure cutoff is open all the way. Check for scale or corrosion at entry and on valve.

cracked the bowl or the tank through the injudicious use of a wrench; or maybe you have noticed that the commode leaks around the bottom. All of these situations call for the removal of the old toilet and either its reinstallation after repair or the installation of a new fixture.

Although there is nothing terribly difficult in removing and replacing a toilet, many communities prohibit — as stated in the official plumbing code — anyone but a licensed plumber from doing the job. Check the code before you even take the first step. If the code does not prohibit do-it-yourself toilet replacement — and if you feel sufficiently confident in your plumbing skill and knowledge — go right ahead with the following procedures.

As you would expect, the first step in removing a toilet is shutting off the water supply. Turn the cutoff valve to its off position, and then remove all the water in both the tank and the bowl. Tripping the flush handle will eliminate most of the water from the tank, but you will need a sponge to soak up whatever water is left. Bail out the remaining water in the bowl with a small container; you will not be able to get all of it, but get as much out as you can.

If you just need to remove the bowl — leaving the tank in place — look to see whether there is a spud between the bowl and the tank. A spud is a large pipe, found most frequently on older units, that is connected at both ends by a slip nut similar to those that hold a sink trap. The nuts that hold the spud, however, are much larger than those on a sink. You must use a special wrench — a spudwrench — to remove the toilet slip nuts. Once you loosen the slip nuts, slide them onto the spud itself. You should have just enough room now to remove the connecting pipe and, therefore, to remove the bowl while the tank is still suspended from the wall. You can understand how convenient a spud is when you just need to get the bowl out to repair a leak or to get rid of a clog that cannot be removed any other way.

What do you do when there is no spud or when you want to remove the tank as well as the bowl? The older tanks are probably connected to the wall; the newer tanks are most likely supported by the bowl. If the tank portion of your toilet is connected to the wall, remove both the pair of hanger bolts from inside the tank and the pair of bolts at the bottom of the tank that connect the tank to the bowl. If the tank is supported only by the bowl, then the latter pair of bolts is all that need concern you. Be sure to keep the rubber gaskets that are under all the bolts; you will need them if you intend to reinstall the tank. With the tank taken out from either its wall mounting or bowl support, you are ready to work on the bowl itself.

Take off the toilet seat; otherwise, it will just get in your way as you work on the bowl. Now, take off the caps over the hold-down bolts at the base of the bowl. You will find either two or four of these caps, most of which are made of ceramic to match the bowl. Some types of caps are held on by compound and can be pried off with a putty knife, while others are threaded and screw on and off. If you do not know which kind of caps are on your toilet, wrap the caps with masking tape and try to unscrew them. If they do not come off, then you know that you must pry them off; brush away the dried compound before proceeding.

Now it is time to remove the nuts or bolts that were hidden under the caps. You may discover that the nuts or bolts are extremely stubborn, but the application of some penetrating oil should make your job much easier. Once the hold-down nuts or bolts are out, there is nothing else holding the bowl. Lift the bowl straight up, but be sure that you have a bucket handy to catch the water that you could not bail out earlier.

With the bowl out and away from the floor, you will find yourself looking down into an uncovered soil pipe. Since you should plug up the hole to prevent any backup of sewer gas, tie a string around an old towel to create a good plug that cannot get away from you. Turn the bowl upside down if you are just trying to remove a clog, but be very careful when you handle either the tank or the bowl. These plumbing fixtures will crack from one bad blow to the porcelain. Therefore, it is a good idea to have an old piece of carpet ready to serve as a work surface on which you can set the toilet.

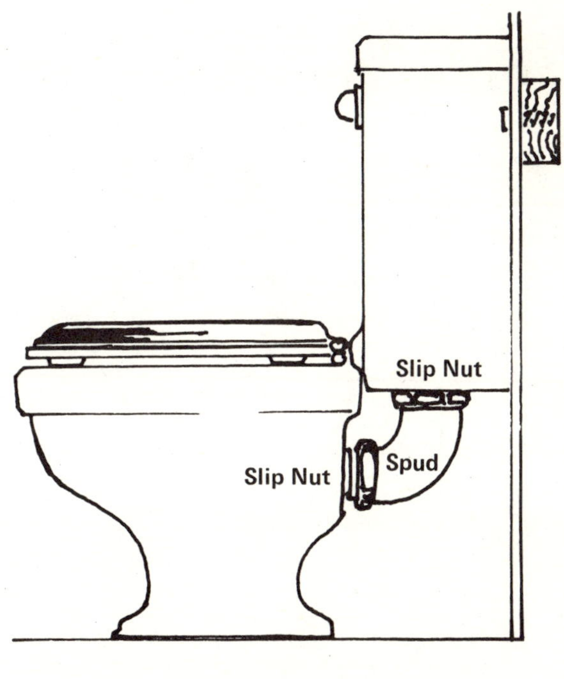

Some old-fashioned toilet bowls are connected to the tank via a large pipe called a spud. The spud may make it possible for you to remove the bowl to clear a clog without also having to remove the toilet tank.

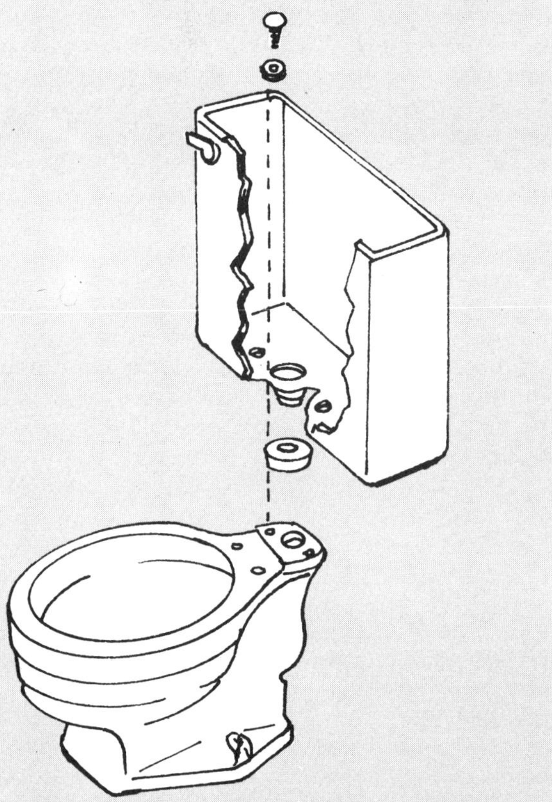

When you remove a bowl-mounted tank, you must take out the two connecting bolts from inside the tank. If you intend to reinstall the same tank, be sure to save the bolts and rubber gaskets, provided they are in good shape.

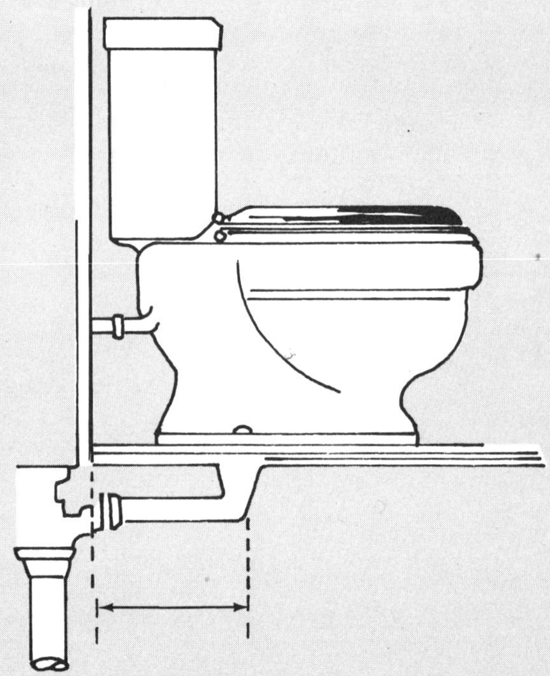

Arrow Is Rough-in

You can replace your old toilet with a different model, but the new unit must fit into the space measured from the wall in back of the toilet to the center of the floor flange. This area is called the rough-in.

After you prepare the bowl base with toilet bowl setting compound, turn the bowl right-side-up and place it down over the flange. Press down firmly, and give the bowl a slight twist to seat the wax ring on the flange.

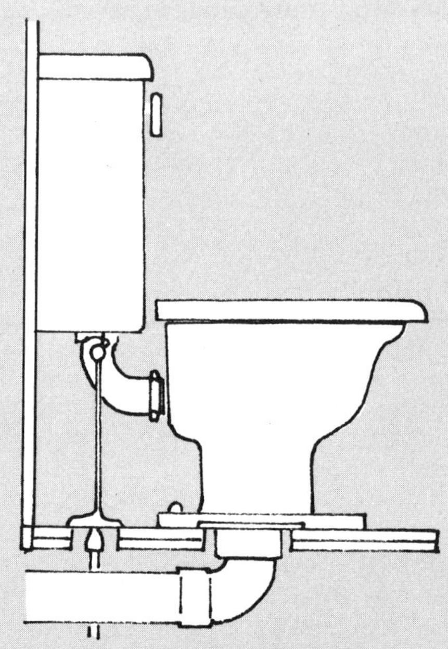

If your toilet has a wall-mounted tank, there are two big-headed hanger bolts that hold the tank to the wall. If the tank and bowl are connected by a curved spud, there are two slip nuts to hold the spud in place.

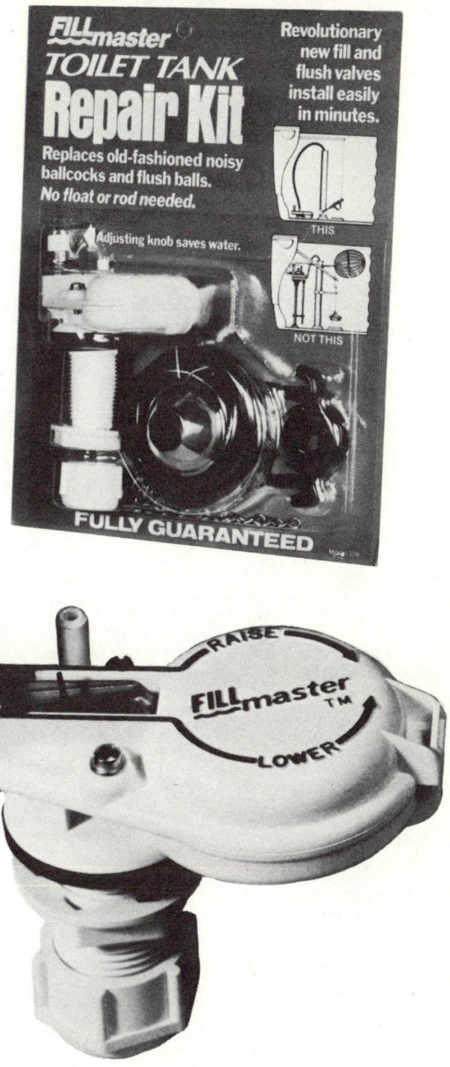

The Fillmaster Fill Valve, a replacement for the standard ball cock assembly, eliminates the float ball and float arm. The unit's diaphragm-powered valve senses the water level from its position at the bottom of the tank.

As long as you are going to all the trouble of removing the toilet from its moorings, you might as well take the opportunity to consider any other alterations that might be needed to make your toilet function better than it does. For example, if the ball cock assembly has been giving you fits, now is the time to put in a new one — especially since you have already done a great deal of the work in disconnecting the tank for removal.

Similarly, this would be a good time to install a separate water cutoff for the toilet if there is not such a cutoff at present. The fixture cutoff allows you to keep the water supply running everywhere else while you work on the toilet. The installation is relatively easy, and most hardware stores carry a kit that contains all the parts you need to add a cutoff to your supply pipe.

You can replace your old toilet with a more modern fixture, but you must make sure that your new unit will fit into the space between the outlet and the wall. This space is called the "rough-in," and it is measured from the wall in back of the toilet to the center of the floor flange. You can install a smaller unit — one that leaves a space between it and the wall — but you cannot put a larger toilet in the smaller space occupied by the former fixture.

Putting in a new toilet and reinstalling the old one are done in the same way. Scrape away all the old putty (or other sealing material) from both the bottom of the bowl and the metal floor flange. Inspect the flange and the bolts that come up from the flange. If anything is amiss — the flange worn away or the bolts stripped — replace the part or parts before you go any further. All of these parts are inexpensive, and it is far better to replace them than to try to get by with parts of doubtful quality.

The next step is to put down a new sealer ring; the best and easiest type to install is the wax toilet bowl gasket. Put the wax ring in place on the bottom of the bowl while the fixture is upside down. You will notice that there is a plastic sleeve in the wax; this sleeve faces up since it will go into the soil pipe. Now, apply a uniform layer of toilet bowl setting compound, about 1/8 of an inch thick, around the base of the bowl. You can buy the compound — sort of a caulking — at most hardware stores and at all plumbing supply dealers.

Turn the bowl right-side-up and place it down over the flange, guiding the bolts in place. Press down firmly, and give the bowl a slight twist to make certain that the wax ring seats properly against the flange. It is, of course, of utmost importance that the bowl be level. Either press down on any higher portion or insert wedges (you can hide them with toilet bowl setting compound) under any lower portion of the bowl to even it. Whatever you do, however, make sure that you do nothing that would affect the seal of the wax ring.

Once you get the bowl positioned properly, you can install the nuts to hold the bowl to the floor. Remember: Do not overtighten the nuts. If you do, you can crack the fixture, and hand-tightening is all that is required anyway. Coat the hold-down nuts and bolts with toilet bowl setting compound and reinstall the caps.

That completes the installation of the bowl; now you are ready to attach the tank. If the tank and bowl are connected with a spud, apply pipe joint compound to the threads of the spud's slip nuts and tighten them in place. Rebolt a wall-mounted tank, or reinstall the bolts and washers that connect a bowl-supported tank. Naturally, before you put them back, make sure that all the washers, gaskets, and bolts are still in sound condition. When you follow all these procedures correctly and then you turn the water back on, you should find that your toilet troubles are ended.

Chapter Four

Tubs & Sinks

Many people who run to call a plumber when a pipe springs a leak or a toilet gets clogged, draw the line at calling in professional help when it comes to tubs and sinks. They feel that it takes no great expertise to stop a faucet from dripping or to clear a slow drain — and they are right. Hopefully, the other sections of this book will encourage you to tackle pipe problems and toilet troubles as well, but certainly every homeowner and apartment-dweller should be able to maintain and repair tubs and sinks without paying for a plumber.

TOOLS FOR TUBS AND SINKS

Most likely, you already possess a great many of the necessary tools for working on tubs and sinks. For the most part, they are tools that can be used for many other do-it-yourself projects besides plumbing chores, or they are tools that are so well-known that no detailed explanation of them is required here. For example, a must for every household is the plumber's friend — also called a plunger or a force cup. The best models have long handles, and their suction cups are large enough to cover different types of drain openings.

Plumber's snakes or augers come in various lengths; you should possess both a short one and a long one. If you plan to buy a plumber's snake, look for one that comes encased in a metal housing — it is far less messy to use. A plumber's tape or ribbon is very similar to a long clock spring.

You will need a wrench for most faucet repairs and for various other connections. A medium-sized adjustable wrench is a fine tool to have around because it can be used on nuts of many different sizes; with one adjustable wrench you can handle the same tasks that would otherwise necessitate a complete set of open-end wrenches. Open-end wrenches, however, do have the advantage over an adjustable wrench in providing a secure grip on the nut. Of course, you can frequently substitute a pair of long-handled pliers for either kind of wrench.

A basin wrench allows you to get up in tight spots under sinks and basins. The jaws of a basin wrench not only adjust to accommodate nuts of different sizes, but they also flip over to the opposite side so that you can keep turning without first

Tub and Sink Tools

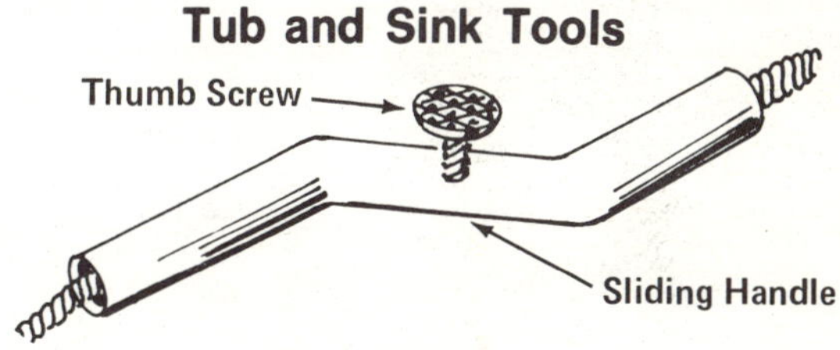

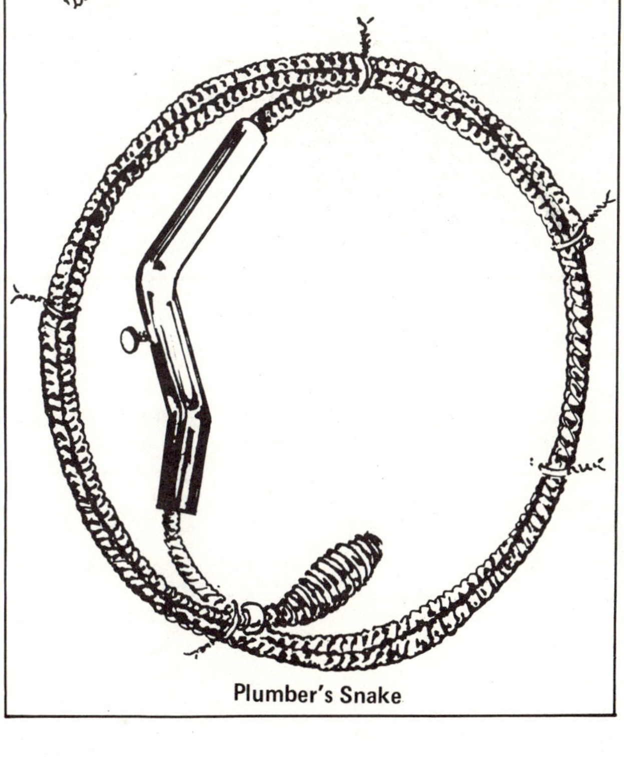

Plumber's Snake

Tub & Sink Tools

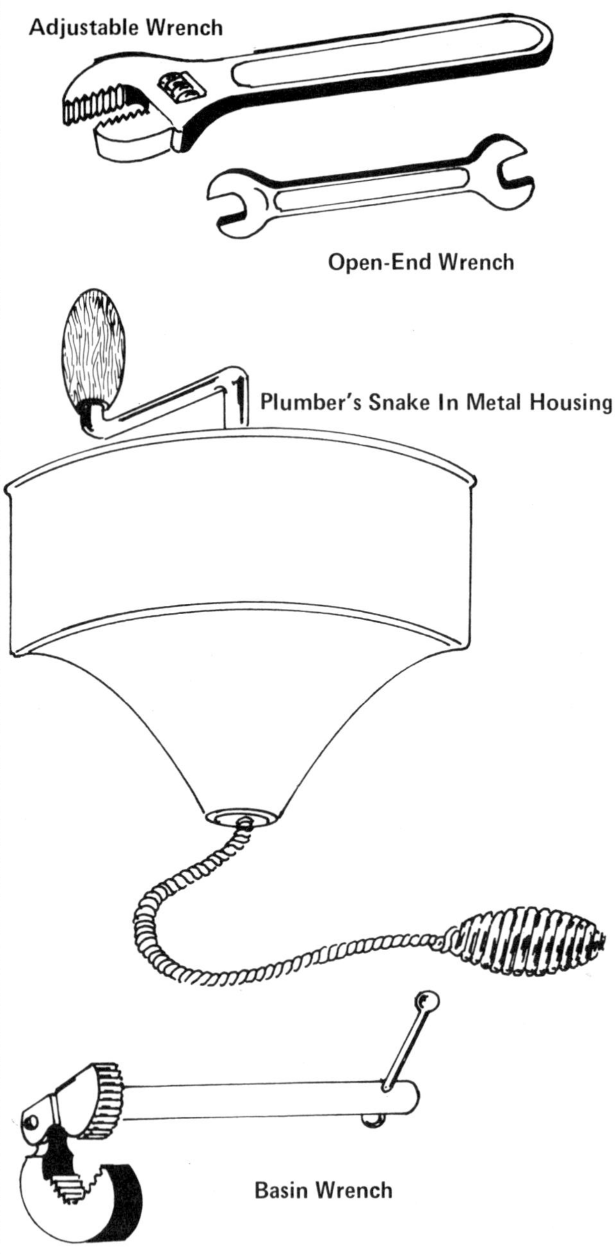

removing the wrench. If you plan to work on tub and shower fixtures, then there is a good chance that you will need a socket wrench set to remove recessed packing nuts. If you do not already own a socket wrench set, it does not pay to buy one just for this job; but you can find many uses for a socket wrench set besides do-it-yourself plumbing chores.

ELIMINATING FAUCET DRIPS

Although the dripping faucet constitutes the most common plumbing problem — and one of the easiest to repair — many people try to ignore it, and they leave the dripping faucet unrepaired. They think, apparently, that a tiny drip is not worth the effort to fix, but those people would be amazed to find out how much that tiny drip costs. A steady drip can add up to so many gallons over a year's time that it can wind up costing about $50 in wasted water. Multiply that figure by the number of such faucet thieves in your home, and you can calculate how much of your money is literally going down the drain. If that is not enough to convince you, just think about the drip that comes from the hot water side; in that case, you are paying to heat the water before you waste it.

A drip is caused by seepage. Remember that the fresh water supply enters your house or apartment under pressure. Therefore, there must be a watertight seal holding back the incoming water when the faucet handle is in the off position. That seal is created by a washer pressed tightly against the faucet seat, holding the water there. Obviously, when the washer or the seat is not functioning properly, a little water can seep through and drip out the faucet spout. To stop the drip, all you need do is replace the washer or repair the seat.

Naturally, the first thing to do when fixing a faucet drip is to turn off the water supply. You should be able to turn off the supply at a nearby cutoff, but if your house or apartment is not equipped with cutoffs for individual fixtures, then you must go to the main cutoff and stop the entire water supply throughout your home.

With the water flow stopped, you can start to disassemble the faucet. No matter what the faucet looks like, if it has separate handles for hot and cold water, it operates according to certain basic principles. The first thing to do is to take off the faucet handle, which is held to the main body of the faucet by a tiny screw — either on the top or at the back of the handle. You may not be able to see a top-mounted screw at first because it is hidden by the metal or plastic button which designates "Hot" or "Cold." These buttons usually snap out, although some are threaded. Once you get the button out, you will see the top-mounted handle screw. You can remove a rear-mounted handle screw directly, but if you find that either it or a top-mounted screw is exceptionally difficult to turn, use a little penetrating oil to free it.

Slide the handle off and look at the faucet assembly. You will see a packing nut (sometimes called the bonnet); remove the nut with either a large pair of channel lock pliers or an adjustable wrench, but take special care not to scar the metal. It is a good idea to wrap tape around the packing nut to protect it from the chewing of your pliers or wrench. Once you get the packing nut off, you can twist out the spindle (sometimes called the stem) by turning it in the same direction you would turn on the tap.

You can see the washer at the base of the spindle, but to remove it you must first take out the brass screw that holds it. That may be very difficult,

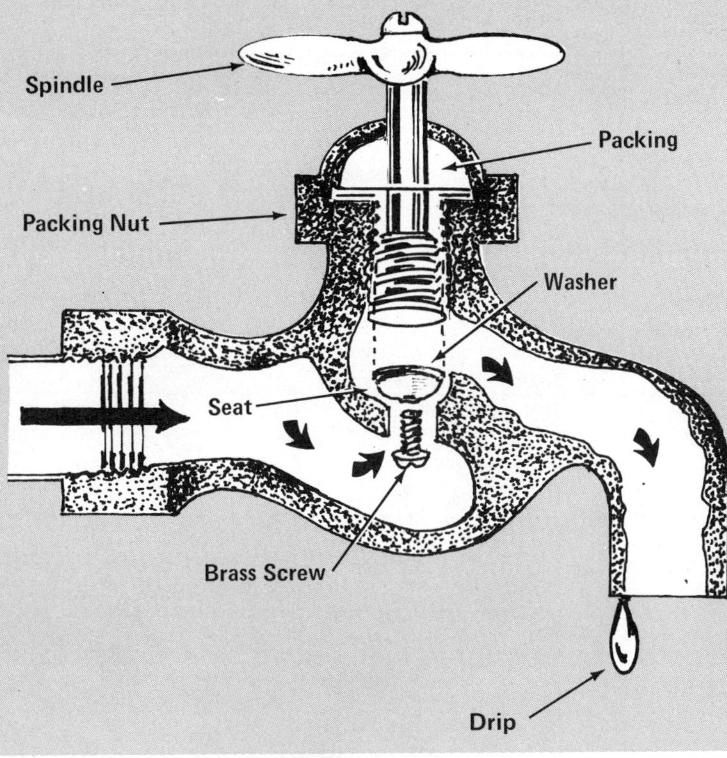

A dripping faucet is the most common plumbing problem and one of the easiest to fix. The drip is caused by seepage when the faucet seat and spindle fail to make a watertight seal against the pressure of the incoming supply. You can usually stop the drip merely by replacing the washer, although you may have to repair or replace the faucet seat as well.

but some penetrating oil should make a stubborn brass screw much easier to remove. If and when you do get the screw out, examine it to see whether it needs replacement along with the washer. If you cannot get the brass screw out, do not despair; you can buy a whole new spindle at very little expense.

Look for the size on the old washer. It is absolutely essential that you put on a replacement washer that is exactly the right size. Washers that "almost fit" will "almost stop the drip." Also note whether the old washer is beveled or flat, and get an exact duplicate for replacement. If you cannot make out the exact size, then take the whole spindle with you to the hardware store. Of course, you can buy a big assortment pack of washers that contains just about every size and shape, but usually you can use only three or four out of the pack and wind up wasting the rest.

Some washers do not work on a hot water tap. A washer designed only for cold water expands greatly when it gets hot, thereby closing the opening and slowing the flow of hot water. Perhaps you have

The first procedure in eliminating a faucet drip is to remove the faucet handle. The top-mounted screw which holds the handle to the main body of the faucet may, however, be hidden by a metal or plastic button which designates "Hot" or "Cold." Either unscrew or pull off the button (depending on the type of button), and take out the handle screw.

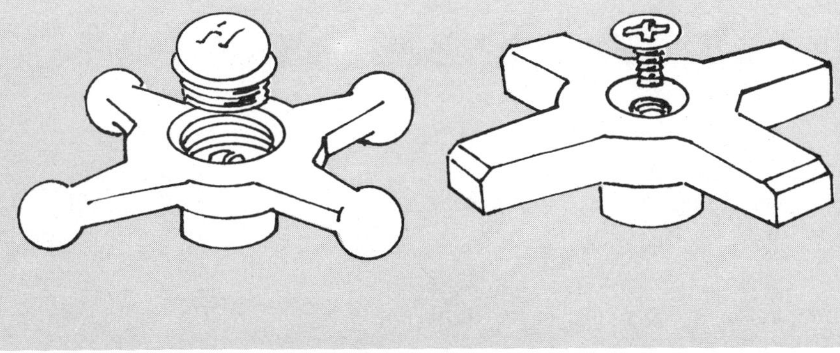

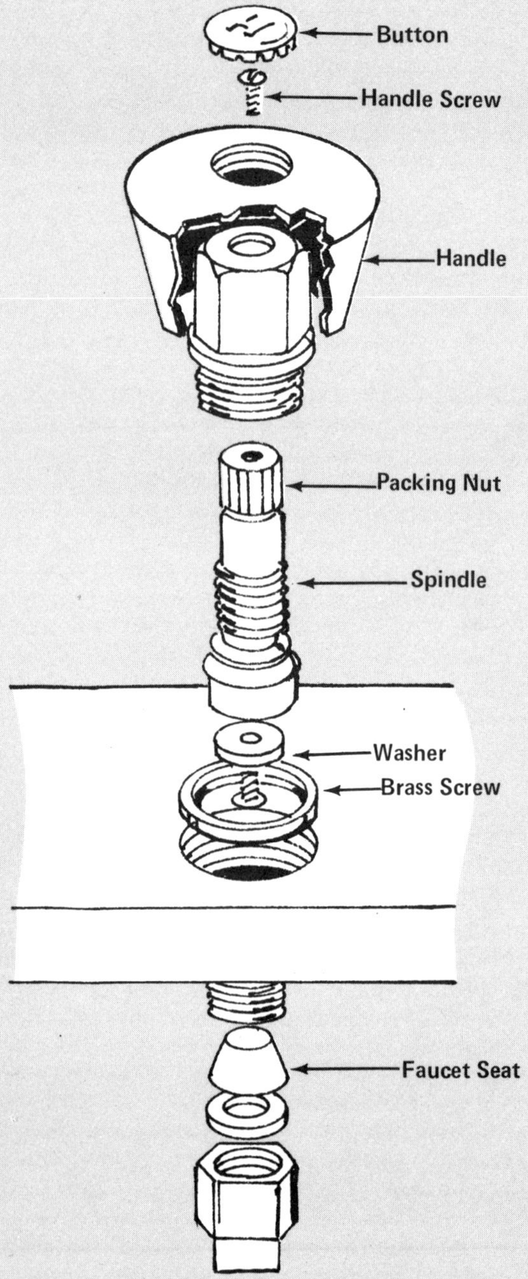

After you remove the handle screw and the handle, you will see a packing nut. Remove the nut with an adjustable wrench, and twist out the spindle by turning it the same way you would turn on the tap.

experienced a hot water tap that works fine until the water gets quite hot; then it slows down to a trickle. Be sure to tell your hardware dealer whether the replacement washer you need is for the hot side or the cold side. Some washers will work for either.

Fasten the new washer to the spindle, and reinstall the spindle back into the faucet (turn the spindle the same direction that you would turn the faucet handle to stop the water flow). With the spindle in place, you can put the packing nut back on, but be careful not to scar the metal. Once you screw on the handle and replace the caps, your faucet is completely reassembled. Turn the water supply back on, and you should find that your days of wasting water (and money!) down the drain are over.

Sometimes, however, you replace the washer and then still find a drip. A drip that exists despite a new washer indicates that there is something wrong with the faucet seat. Perhaps a defective washer at some point in the past allowed the metal spindle to grind against the seat and leave it uneven, or else the chemicals in the water have built up a residue that now prevents the washer from fitting tightly against the seat.

What do you do to repair a bad faucet seat? You can purchase an inexpensive reseating tool that will even out a worn seat, but be careful not to use the tool too long or with too much force because the faucet is made of soft metal and you can grind too much of it away before you know it. The other thing you can do is to replace the seat — a necessity if you grind it down too far with a reseating tool. Removal of the old seat is fairly simple if you have the right tool; just insert the tool into the seat and turn it counterclockwise. Once you get the old seat out, be sure that the replacement seat you buy is an exact duplicate. You may, on occasion, come upon a seat that is impossible to remove; in such a case, you can insert a seat sleeve that slides into place in the old seat and provides a tight seal.

FAUCET LEAKS

A drip occurs when the faucet is turned off; a faucet leak occurs with the water running. If you see water coming out around the handle, then you have a faucet leak. The first thing to do is to make sure that the packing nut is tight (but be careful not to scar it). If the nut is not causing the leak, then the second repair procedure is to replace the packing. Faucet packing can be a solid piece; it can be an "O" ring; or it can resemble spaghetti that is wrapped around the spindle under the packing nut. To replace the packing, you must remove the faucet handle and slip both the packing nut and the old packing up off the spindle. Put the new packing on, but before you close up the faucet, smear a light coat of petroleum jelly over the threads of the spindle and over the threads of the packing nut.

Kitchen faucets — the ones where the spout swings from side to side — present a somewhat different situation. These faucets possess an "O" ring (sometimes more than one) that prevents water from oozing out around the spout. If the rubber "O" ring wears out, you will see water at the base of the spout every time you turn on the water. To replace the ring (or rings), remove the threaded coupling nut that holds the spout in place, but be sure to wrap the

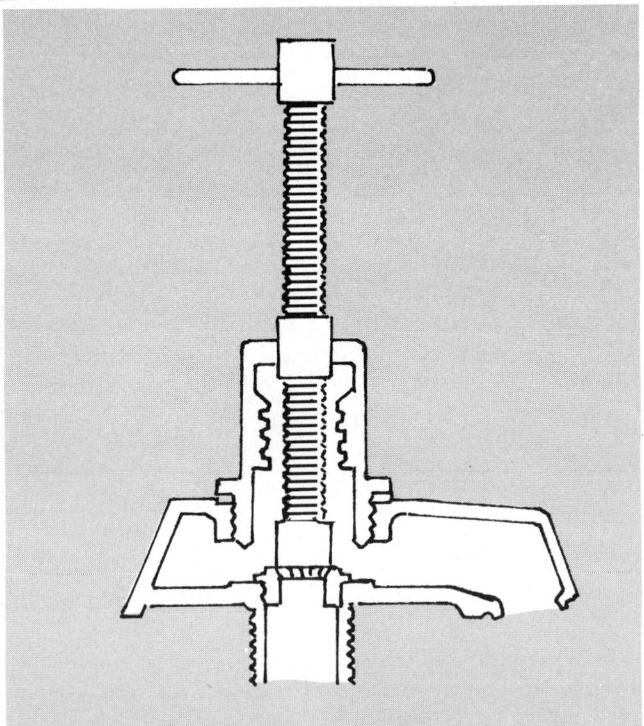

If you need to repair a worn-out faucet seat, you can purchase an inexpensive reseating tool that will even out the seat by grinding away the uneven portions. Be careful not to grind off too much of the metal, however.

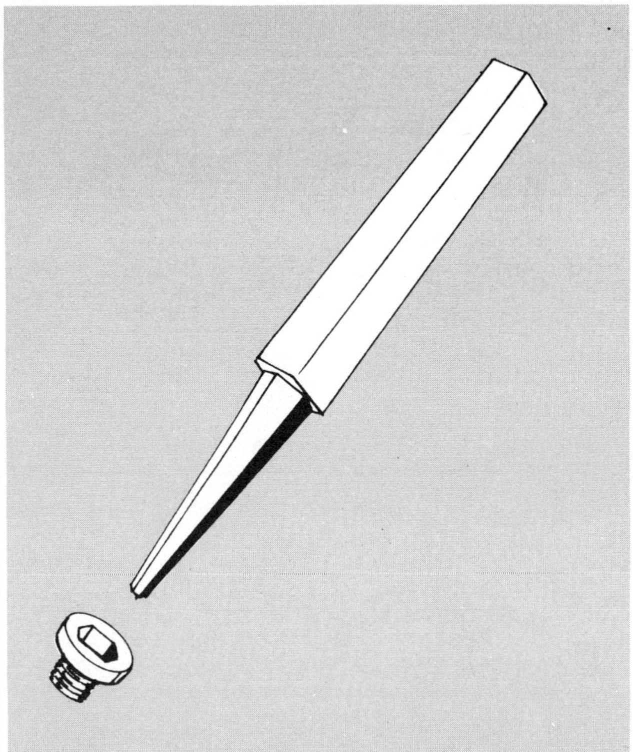

Replacing a faulty faucet seat is simple if you have the right tool (above) for the job. Just insert the tool in the seat and turn it counterclockwise for removal.

Kitchen faucets that have spouts which swing from side to side possess an "O" ring to prevent the water from oozing out around the spout. If you notice water at the base of the spout, replace the "O" ring.

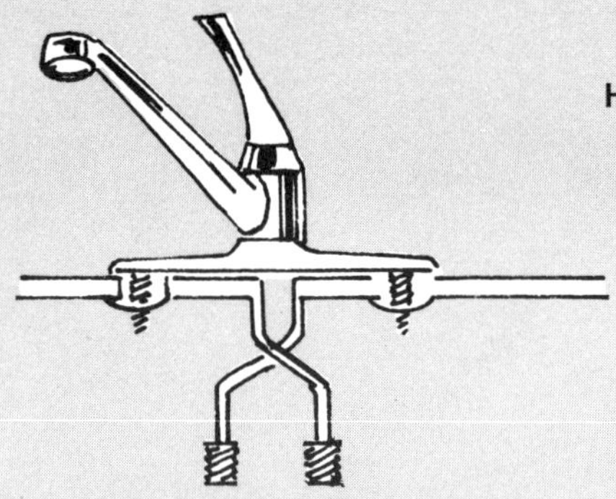

Cross Water Supply Tubes in Faucet

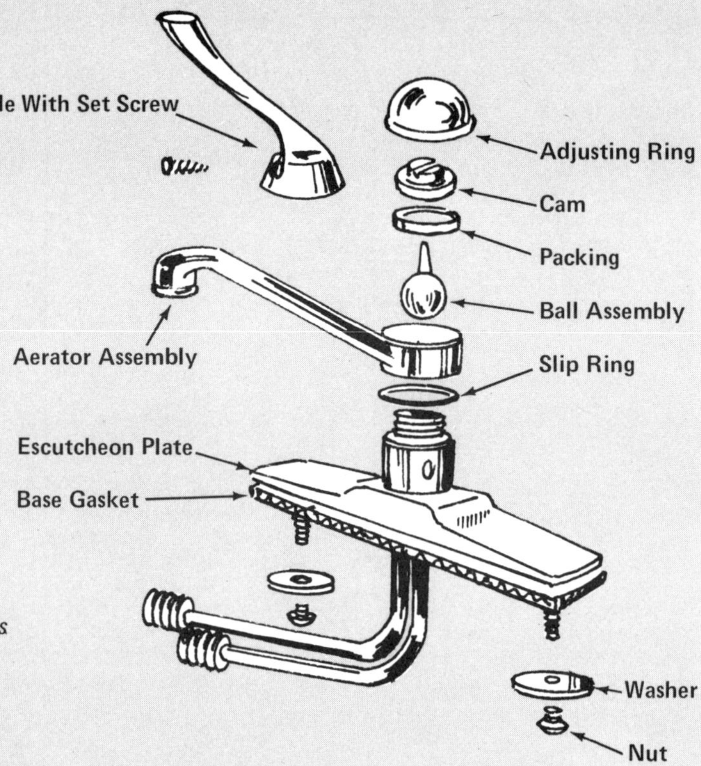

Single-handle faucets are simple to fix, but you must purchase the manufacturer's repair kit for the specific faucet you have. The kit contains the parts you need, as well as detailed instructions and diagrams. If you want to install a single-handle faucet as a replacement, just make sure that the new unit covers the holes left by the old faucet and that you cross the copper tubes on the new faucet before you reconnect the supply lines.

nut with tape to prevent scratching it. With the coupling nut removed you can work the spout up and out of its socket, where you will find the ring(s). Replace the defective ring or rings with new ones of exactly the same size, reassemble the faucet, and you should find that your kitchen faucet works without leaking a drop.

The modern-looking single-lever faucets are simple to fix as well, but there are so many different types that you must purchase a specific repair kit for the faucet you have. Generally, the faucet manufacturer makes repair kits for his products, and includes detailed instructions and diagrams along with the replacement parts. The hardest part about repairing a single-lever faucet may be in tracking down the hardware dealer or plumbing supply store that carries the appropriate kit. Once you have the kit, however, you should find little difficulty in eliminating the leak. Just make sure that the water supply is shut off before disassembling a single-lever faucet (as you would were you working on a standard double-lever faucet), and follow the kit's directions carefully.

SILENCING NOISY FAUCETS

Faucets can scream, whistle, or chatter when you turn them on or off. There are several possible causes for these ear-shattering phenomena. If your house is newly constructed, you probably have pipes that are too small to allow the water to pass properly. Similarly, pipes in older homes can become restricted by the formation of scale, with the same result — a noisy faucet. In either case, you must replace the pipes to get rid of the noise.

Most likely, however, your noisy faucet is caused by a washer that is either of the wrong size or that is not being held securely to the spindle. Just replace the washer, or tighten it, and you should eliminate the noise. If the faucet still makes noise, then check the washer seat; the seat can get partially closed with residue, and the restricted water flow then can whistle or chatter.

A faucet that emits a terrible squealing noise when you turn the handle is one in which the metal threads of the spindle (or stem) are binding against the faucet's threads. Remove the spindle and coat both sets of threads with petroleum jelly. The lubrication should silence the noise and make the faucet easier to turn. Of course, if the spindle threads and faucet threads have become worn away, the resulting play between them causes the vibration and noise in the faucet. You will need more than just lubrication to quiet such a faucet. Install a new spindle and see if the noise vanishes. If not, it means that the faucet threads are worn, and the only solution is a new faucet. Fortunately, the spindle is the part that usually goes first, but even if you must replace the entire faucet, you will find it to be an easy and inexpensive do-it-yourself project.

CHANGING FAUCETS

Changing faucets — either for functional or aesthetic reasons — gets you into a little more work than just changing a washer or putting in a new seat, but it certainly is a feasible project. Fortunately, all of the new faucet units are made for do-it-yourself installation, with easy-to-follow instructions included. A new faucet can work wonders for the appearance of your old fixtures, and — even better — it will eliminate all the leaks, drips, and other problems you had with your former faucet.

Suppose, for example, that you want to replace the old faucet in your kitchen sink with a modern single-handle unit. Just make sure that whatever unit you select will cover the old faucet's holes; generally that is no great problem, but some sinks — especially older ones — can be somewhat strange. If you have an unusual sink, look for one of the adjustable faucet units which are made to fit many types of sinks. Once you select the faucet model that you want, follow these steps to install it properly:

1. Cut off the water supply to both the hot and cold faucets.
2. Disconnect the old faucet from the supply lines. It will probably be a threaded connection, but no matter how it is connected, you should have no trouble removing it.
3. The old unit is probably held in place by nuts underneath the sink. Remove the nuts and take out the old faucet. With the faucet out of the way, clean the area around and under where the new unit will go.
4. Place the gasket in position over the supply tubes and the lugs of the new unit.

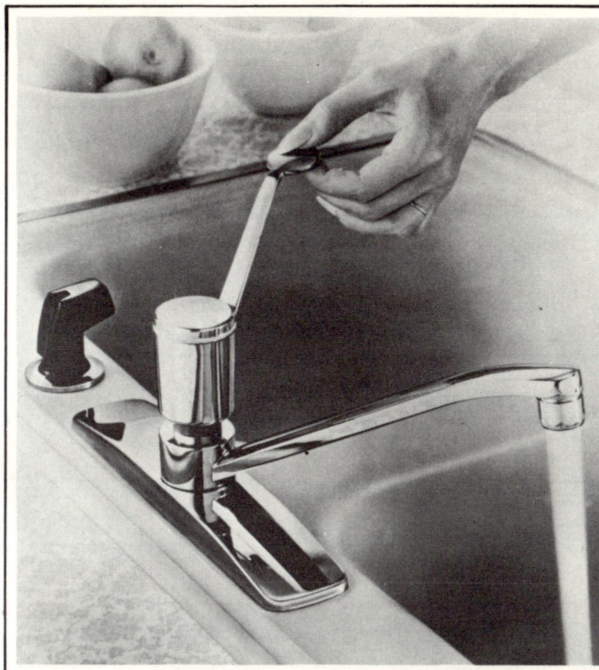

Replacing a double-handle faucet with a single-handle unit is no problem if the new faucet covers the old holes.

Tub and Sink Troubleshooting Chart

PROBLEMS	CAUSES	REPAIRS
Faucet Drips	1. Faulty washer.	1. Replace the washer. For single-handled faucets, install all parts in repair kit.
	2. Improper seat.	2. Use reseating tool to grind seat even, or replace seat.
Hot water slows to trickle	1. Washer expands when hot.	1. Replace with proper nonexpanding washer.
Leaks around handle	1. Packing nut loose.	1. Tighten the packing nut.
	2. Packing not adequate.	2. Replace the packing.
Leaks around spout	1. Faulty O ring.	1. Replace the O ring.
Handle hard to turn; squeals when turned	1. Spindle threads binding against threads in faucet.	1. Lubricate spindle threads with petroleum jelly, or replace spindle.
Drains overflowing	1. Clogged pipes.	1. Use plumber's friend, chemicals, aerosol, or snake to remove blockage.

A push-pull single handle faucet can do wonders for the appearance of your bathtub, but installation can be complicated if there is no access panel to the pipes.

Most of the attractive single-handle faucets are made for easy do-it-yourself installation. Just follow the manufacturer's instructions and diagrams on the package.

5. Feed the tubes down into the center hole, and position the threaded lugs into the outer holes.
6. Go below the sink and install the washers and nuts to secure the unit squarely in place.
7. The copper tubes now need to be bent in a cross pattern. The reason they are crossed is because the hot water comes into the right side, as you face the unit, and the cold from the left. This is the reverse of the way you expect your faucet to operate: cold from the right, hot from the left. Bending must be done very gently; take great care not to kink the tubing.
8. Once the tubes are crossed, you are ready to reconnect the supply lines. Make sure that the cold and hot lines are connected to the correct tubes on the unit. When you attach the lines, use pipe joint compound and be sure to use two wrenches. One must hold the shank of the fitting on the tube, while the other turns the nut on the water supply line. If you fail to hold the tube, you could twist it and stop the flow of water. If you need different sized connectors, an adapter can do the job.

Once you make the connection on both sides, you can turn the water supply back on because your new kitchen faucet unit is completely installed. You can put in bathroom lavatory faucets following essentially the same procedures. Again, be sure that the new faucet will cover the old holes. Shower and tub faucets, however, can be a bit more complicated since the connections are not made under a sink but rather behind a wall. Whoever built your home should have provided an access panel to allow you to get at the pipes without ripping the wall apart. If you find that you must cut into the wall, however, be sure to add an access panel for pipe and faucet repairs.

Actually, once you get to the tub faucet connections behind the wall, the job is no harder than working on your kitchen sink. Just cut off the water, remove the handles on the tub side, and then disconnect the old faucet unit from the back. Unscrew the old shower head pipe from its pipe inside the wall, and do the same thing with the tub spout. Now you are ready to install all the new parts. Just follow the directions that are included with the faucet units, and you can have the job done before you know it.

ADDING A WATER OUTLET

What about those situations in which you would like to have a faucet where none existed

Modern single-handle faucets not only give your sinks and tubs a contemporary appearance, but they also provide you with enhanced convenience and control.

before? For example, you may want to add a drinking fountain to the middle of a pipe that runs up an exterior wall, or you may like the idea of having a water faucet close at hand in your new built-in bar. You probably think that putting in either of these new additions would require a great deal of plumbing expertise — cutting into the pipe and installing special fittings — but you can do these things yourself more easily than you ever thought possible.

There is a secret to adding a water outlet; the secret is a gadget called the saddle tee, which is composed of a curved plate that fits around the back of the pipe and bolts to a front plate that has a threaded opening to accommodate a faucet. The threaded openings come in several sizes for various popular connections, and the plates are made to fit several pipe sizes. The front plate is backed by a washer or gasket, and it has a hole through which the water flows from the pipe.

Installation is simple. Place the saddle tee on the pipe at the point where you would like to have the water outlet. Tighten the bolts to hold the two plates together securely, and then turn off the water supply to that pipe. Drill a hole straight through the front connector plate and on through the front side of the pipe. A hand drill is the best tool to use for this purpose, even though it is somewhat slow. A power drill is much faster, but it can sometimes go through both the front and the back of the pipe before you can stop it. Therefore, use a hand drill with a bit that will make a hole in the pipe as large as the hole in the front connector plate. After you drill the hole, you will find a great many metal filings in the threaded opening of the front plate. Remove the filings by blowing them out with a squeeze bottle; never try to blow the chips out with your breath because you could blow them right into your eyes.

The only thing remaining to be done is the installation of the faucet itself. Apply pipe joint compound to the threads, and then screw the faucet into the threaded connector of the front plate. That is all there is to it; just turn on the water supply, and you have an outlet exactly where you want one.

Saddle tees are extremely versatile plumbing devices. You can purchase saddle tees for either copper or iron pipes and for either hot or cold water. You can put in a saddle tee when you want to connect a branch pipe to a main pipe or to make any other threaded connection to a fixture. For example, you can hook up a dishwasher or a clothes washer with a saddle tee. You would probably need to tap into both hot and cold water pipes, but then you would just thread a faucet connection and a hose to carry water to the appliance.

One of the best saddle tee installations — best because a professional will charge you from $15 on up for doing the same work — is the one that connects the water supply to the ice-making device in your new refrigerator. In fact, for a few dollars you can buy a kit that contains both the saddle tee and the connector for the copper tubing.

If you do decide to make the water connection to the ice-making device a do-it-yourself project — and there is no reason not to do the job yourself — here are some hints to remember. Once you install the copper tubing, bend it so that it cannot vibrate against any other refrigerator components; if the tubing were to strike anything, you would hear some strange noises coming from your new appliance. Keep one other thing in mind: Like any other plumbing fixture, your ice-making device should be equipped with a water cutoff. You may never need to use the cutoff, but without one you would have to shut off the entire water supply system any time you wanted to work on your ice-maker connection.

INSTALLING A CUTOFF

Any plumber who installs a fixture without simultaneously installing a cutoff (or, as it is more properly called, a supply stop valve) is only doing part of a complete job. You will discover the truth of this statement — if you have not discovered it already — the next time a plumbing crisis occurs and you must waste precious time running to your home's main water supply valve.

Fortunately, you can add cutoffs to all your plumbing fixtures at relatively modest cost. Whether you work on a sink, basin, or appliance, the cutoff principle is the same. You must remove the section

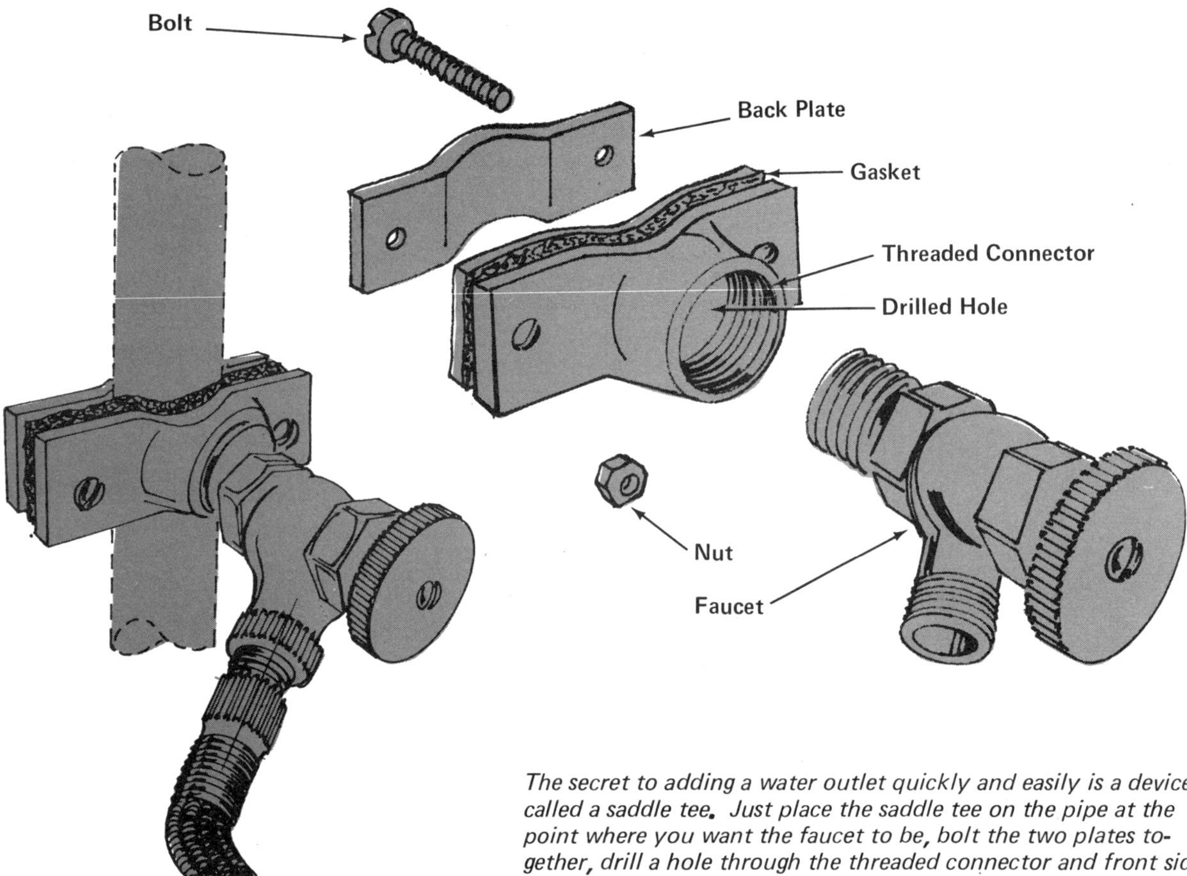

The secret to adding a water outlet quickly and easily is a device called a saddle tee. Just place the saddle tee on the pipe at the point where you want the faucet to be, bolt the two plates together, drill a hole through the threaded connector and front side of the pipe, and attach the faucet.

of pipe that runs to the fixture from under the floor or behind the wall, connect the cutoff to the piece of pipe sticking out (called the stub-out), and then connect the cutoff to the fixture with a flexible chrome-plated copper tube (called a speedee).

The kind of stub-out that you have dictates the kind of fittings that you need. A threaded pipe naturally requires a threaded cutoff; a copper stub-out takes an adapter which you must sweat solder at one end, but which is threaded at the other end for a threaded cutoff. If the stub-out comes from the wall, you need an angled stop to turn the water flow at a right angle; if the stub-out comes from the floor, however, you need a straight cutoff.

The speedee makes installing the final connections easy because it saves you from cutting or piecing pipe together to join the stop to the fixture. Speedees come in three different styles for three different purposes: the one with the flat flange end is for toilet tank connections; the one with a bayonet end is for lavatories; and the speedee with the threaded tip is for kitchen sinks. Unless you buy a kit which contains a stop and speedee already joined, you must join them yourself with a flared connection, a compression fitting, or a compression ring fitting. While you save money making the connections yourself, the extra expense of a kit is money well spent in terms of the work it saves you. In addition to the stop and the speedee, the kit also provides all the connectors, rings, and other items — including complete instructions — you need to do the job right.

REMODELING OR MODERNIZING SINKS AND TUBS

There is no reason why you have to live with plumbing fixtures that are either inefficient or unattractive or both. Do you have a bathtub that is perched up so high on legs that you feel as though you must pole vault to get into it? Or perhaps you have always wanted a double sink in the kitchen instead of a single unit. There may even be places in your home where you want to add new plumbing facilities in areas that had none before. A half bath in the basement, a sink in the workshop, or a guest powder room in what is now a large closet — all these ideas make your home more convenient and, of course, more valuable.

The best part, naturally, is that you can remodel or modernize your plumbing fixtures your-

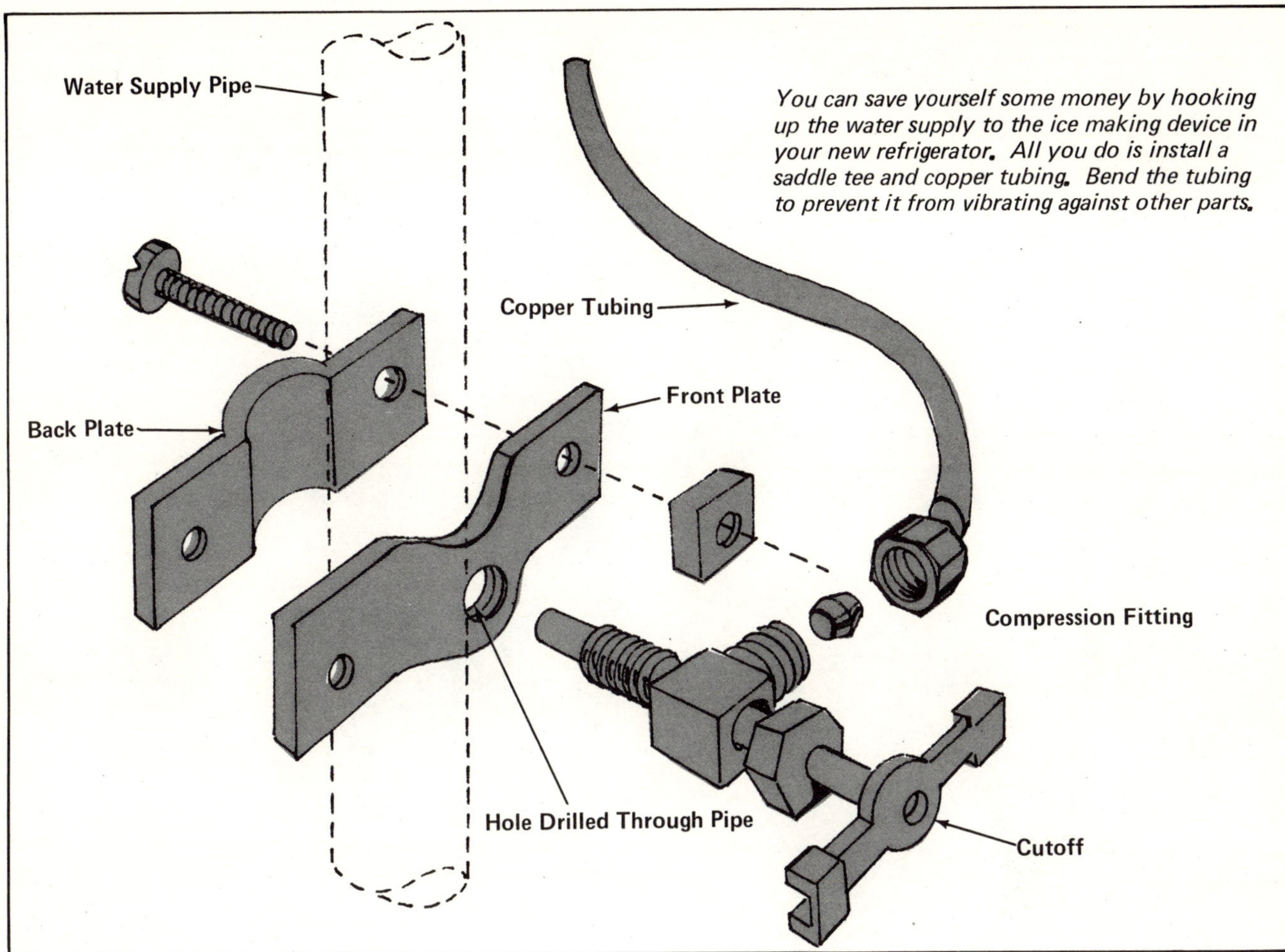

You can save yourself some money by hooking up the water supply to the ice making device in your new refrigerator. All you do is install a saddle tee and copper tubing. Bend the tubing to prevent it from vibrating against other parts.

self, thereby saving a great deal of money. Since the idea is to improve your home, however, do not scrimp when it comes to buying the new fixtures. This is not the kind of change that you make every day, and you owe it to yourself — for whatever labor is involved — to put in the very best you can afford. The relatively small amount extra that you must pay initially to purchase top quality fixtures shrinks to insignificance over the long period of time you will be using the lavatory, tub, or shower.

Remodeling or modernizing — in contrast to a great many other plumbing projects — allows you to take your time and plan everything carefully. Most likely, you have lived with the old fixtures for some time, and a few more weeks will not make any difference. This is no emergency situation that demands a snap decision. Take your time and investigate all the new types of fixtures that can enhance the efficiency as well as the beauty of your home.

There is one other thing that you must investigate before you start replacing old fixtures or adding new ones: the local plumbing code. Almost every community has a plumbing code which governs the materials and procedures required for all plumbing work, and some codes specify that only a licensed plumber can perform certain plumbing installations. Such a provision may compel you to contract with a plumber for part of the job while doing the rest yourself. In any case, familiarize yourself thoroughly with all the code's provisions, and do only the work which it allows you to do legally.

Usually, the building inspector's office in your city has copies of the code. If such an office is not available, check city, county, or state offices until you find the code. Then, check with the same people who have copies of the code to see about a plumbing permit. Although the permit may cost you a few dollars, it is a smart idea to get one. The authority will inspect your plumbing projects to insure that your plans meet the requirements of the code, and an inspector will check afterwards to see that the work was done properly. Plumbing must be done right; otherwise, the health of your family and the well-being of your entire neighborhood could be endangered. Making certain that your plumbing project is done according to the code is sure to pay off in the long run.

Suppose you want to install a new bathroom lavatory; you should have little trouble with the code

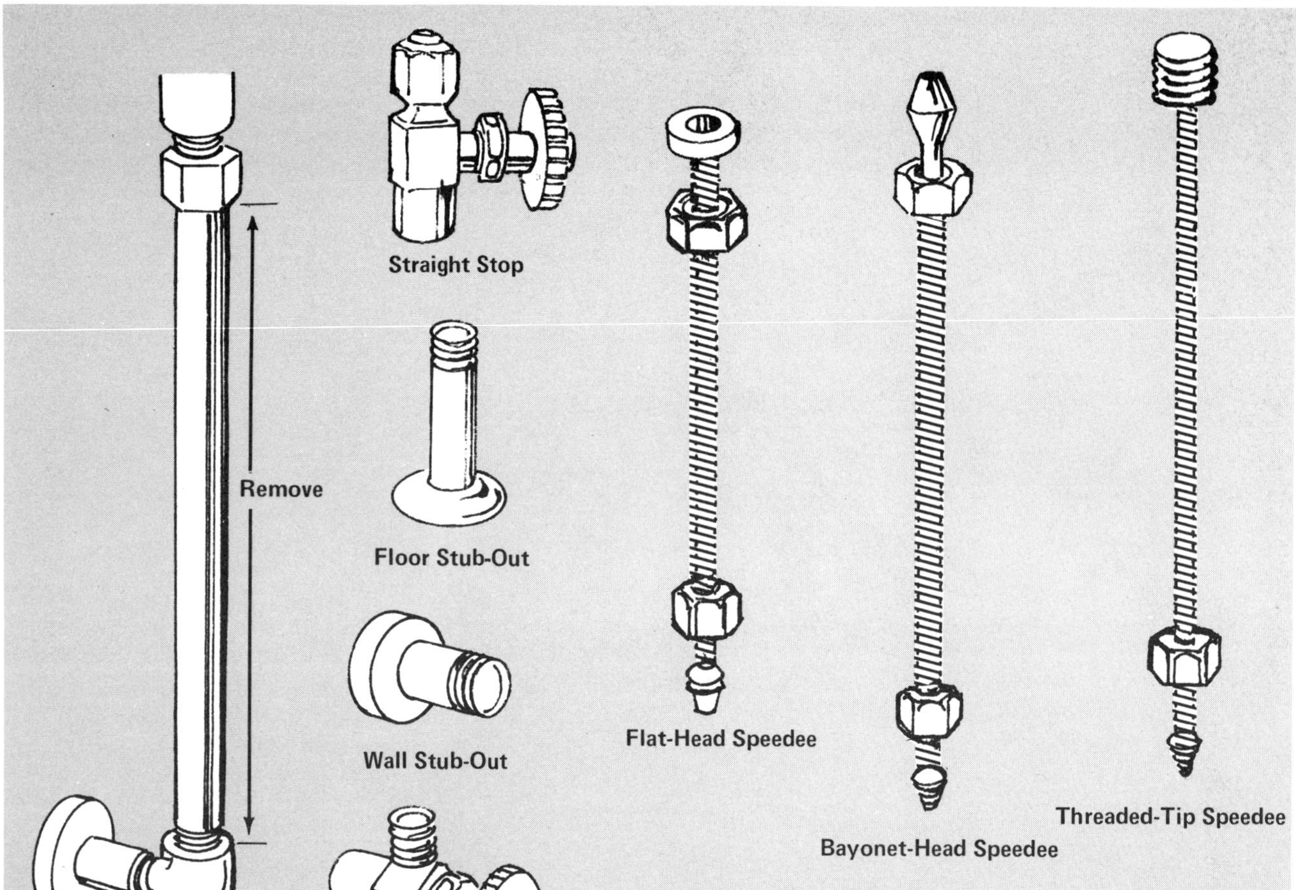

To install a cutoff, you must remove the supply pipe that runs to the fixture. The part remaining is called a stub-out. If the stub-out comes from the wall, install an angled stop; if it comes from the floor, install a straight stop. The speedee runs from the stop to the fixture.

because all you are doing is hooking up a new fixture to the existing pipes. Chances are that you will not even use new adapters or fittings. If your old plumbing is ugly, you can hide it easily with a cabinet-type lavatory that also gives you some additional storage space. For a wall-hung, leg-supported, or pedestal-type lavatory, however, you can replace any unattractive old fittings with shiny new chrome ones.

A wall-hung lavatory slips into place on a wall bracket, which is fastened to a wooden cross-piece. Nail the crosspiece to at least two studs (it will look better if it is flush with the wall), and screw the bracket on after positioning it with a carpenter's level. The basin hangs in the bracket while you make the plumbing hookups underneath.

A kitchen sink replacement may be a bit more involved. If you want to put in a larger sink — a double unit, for example — where a smaller single unit now exists, then you may be faced with changing the cabinets as well. From the standpoint of the plumbing task, however, replacing a kitchen sink is just as simple as any lavatory replacement. The new sink fixtures will probably be compatible with the old connections, and it is doubtful that you will need any additional adapters or fittings.

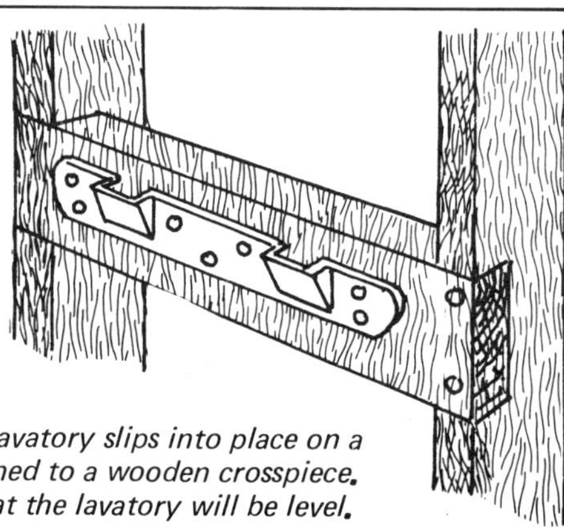

A wall-hung lavatory slips into place on a bracket fastened to a wooden crosspiece. Make sure that the lavatory will be level.

You can install a contemporary-styled bathtub yourself if the existing plumbing rough-in is suitable.

Where you really get into carpentry work with remodeling or modernizing plumbing fixtures is with new tubs and showers. A new tub that fits into the area occupied by the old one may just involve plumbing hookups, but if you opt for a tub that is bigger than the allotted space, you must move some walls — or at least part of one wall. Bathtubs are made in several different widths and lengths, but your choice is limited by the existing rough-in (the location of the supply and drain lines in relation to the distances from the wall and from the floor). You must choose a new tub in the size and style that is compatible with the old pipes.

Suppose, for example, that you want to put in a combination tub and shower where your old tub on its high legs now sits. Here is what you must do. Since the old tub was not closed in at all, you must install walls on each end and make the side wall waterproof. First frame the wall at the end where the pipes are located because you want to make sure that you hide the pipes that were out in the open on the old-fashioned tub. Before you frame the wall at the other end, check to see if the new tub needs supports running around the three walled-in sides. If so, install the supports at the height indicated. Since it is best to nail the supports directly to the studs, remove the plaster or tile from the surrounding walls and put in the supports. Then you can frame the wall at the rear of the tub.

Now you are ready to connect the existing pipes and to add the pipe that goes up to the shower outlet. Start with the existing hot and cold water pipes and add whatever adapters, extensions, or offsets it takes to bring these two lines up to where they will connect to the new tub/shower combination unit. The pipe for the shower requires at least one crossbrace in the framing and several U clips for steadiness in its vertical run.

To complete the adaptation of the old rough-in to the requirements of the new tub/shower combination, you need to notch a crosspiece to hold the faucet unit and its stub-outs, and you must alter the drain pipe to fit both the overflow and the drain openings of the new tub. With the rough-in finished,

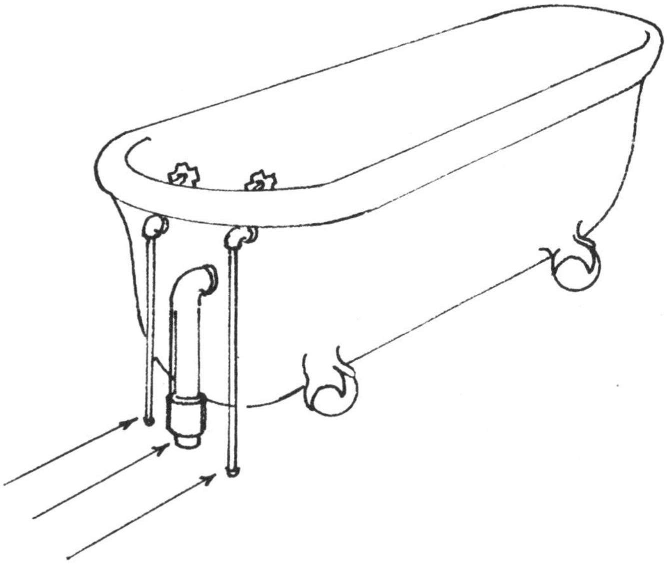

When you replace an old-fashioned tub, take out everything except the hot and cold water supply pipes and the drain pipe.

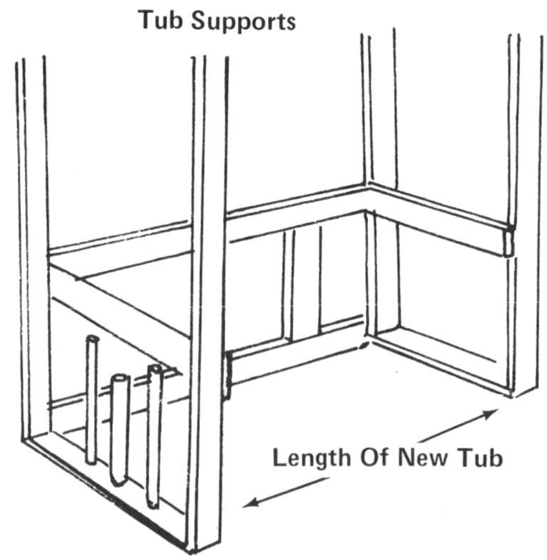

Since most old bathtubs are not closed in, you must provide walls on each end of your new tub as well as waterproof the side wall.

you are ready to position the tub on its supports and fasten it down according to whatever method of securing the tub that the manufacturer recommends.

Now you can finish the three surrounding walls. You must make them waterproof and, of course, you want them to look as good as your new tub. There is one other factor to keep in mind as you waterproof and beautify the walls, however, and that is a functional consideration called an access panel. Before you complete the back side of the new wall at the head of the tub, you would be very smart to install an access panel that will provide easy entry to pipes in need of repair.

With the walls completed, all you have to do is add the handles, plates, spout, and shower head. Your new tub/shower combination is installed, and your days of leaping up into that old-fashioned leg-mounted fixture are over at last.

TUB AND SINK STOPPERS

You flip up a lever at the end of the tub or you raise the lifter handle between two faucet handles on the basin, and in both cases the stopper closes the drain. Who gives these stoppers a second thought? Yet, when a plumbing problem occurs and you need to get into the drain, then you give that stopper plenty of thought because that is when you must get the tub or sink stopper out of your way.

The most common lavatory basin stopper is the pop-up hollow cylinder that is attached to a rod that sticks out into the drain pipe. When you pull on the lifter, a lever pivots a ball that causes the rod to go down and pull the stopper down with it. Since the top of the cylinder is closed, the water stays in the basin when the cylinder is down. Thus, with the cylinder down, the drain is closed.

A pop-up hollow-cylinder stopper is easy to remove. Just apply a slight downward pressure, twist (most of them twist clockwise, but some go the other way), and pull out. Remove the basin stopper at least once a month, and clean out the hair and whatever else is caught in it. In addition, corrosion or dirt can collect around the lip of the stopper, causing the basin not to hold water. You may not even be able to see any residue, but removal and cleaning of the stopper should stop the seepage.

A popular tub stopper is called the trip-lever system in which there is no visible stopper at all. When you flip the lever, the trip-lever system closes the drain by lowering a metal plug into a position that blocks the water flow. The lever itself is mounted on an open plate that is designed to accommodate any tub overflow. To remove a trip-lever system stopper, all you have to do is take out the two screws in the plate and lift the entire assembly straight out; it goes back in the same way.

There are two types of sink stoppers that connect directly to the strainer. One employs a screw nut to hold the unit, while the other utilizes a retainer with screws that must be tightened to secure the unit. To remove either type of stopper, you must unhook the trap from the tailpiece underneath, remove the large nut that holds the drain flange inside the sink, and tap the tailpiece upward until it pops loose. Scrape out all the old putty from where the flange was in the sink, and then put in fresh putty around the opening before installing a new stopper.

If you are putting in a unit that has a screw nut, put the flange in the hole, and — from the bottom — place the rubber gasket and metal washer in position, followed by the threaded nut. Tighten the nut until the strainer is watertight. You put back a retainer unit in just about the same way, except that you tighten the three retainer screws against the friction ring until the strainer is watertight. Then, with either type of unit, you connect the tailpiece with a slip nut, reattach the trap, remove any excess putty, and your new strainer/stopper is ready for service.

CLEARING CLOGGED DRAINS

Of the two most common plumbing problems — dripping faucets and clogged drains — the latter is undoubtedly the one that receives the most attention. Most people fail to realize how much a drip costs them, but they all know the inconvenience as well as the mess that accompanies a sluggish drain. Even so, many people wait until the drain situation becomes dire — that is, when no water runs out — before they take corrective action.

And when they do finally start their corrective action they usually fail. That is a result of their not knowing how to use one of the best drain-clearing tools of them all: the plumber's friend or, as it is often called, the rubber plunger. Working a plunger requires no special training or expertise, but if it is to do its job properly, you must know a few basic facts. Otherwise, you will follow the familiar syndrome: pump up and down two or three times, step back to see what happens, and then go on to other drain-clearing devices — or even call in a plumber — when the water fails to whoosh out the drain.

Here are some of the facts you must know if you expect to clear a clogged drain with a plunger. First, the suction cup must be large enough to cover the drain opening completely. Second, the water in the sink or tub must cover at least a couple of inches from the bottom of the plunger. Third — and undoubtedly the most neglected aspect of using a plumber's friend — you must block off all the other outlets between the drain and the blockage. If you

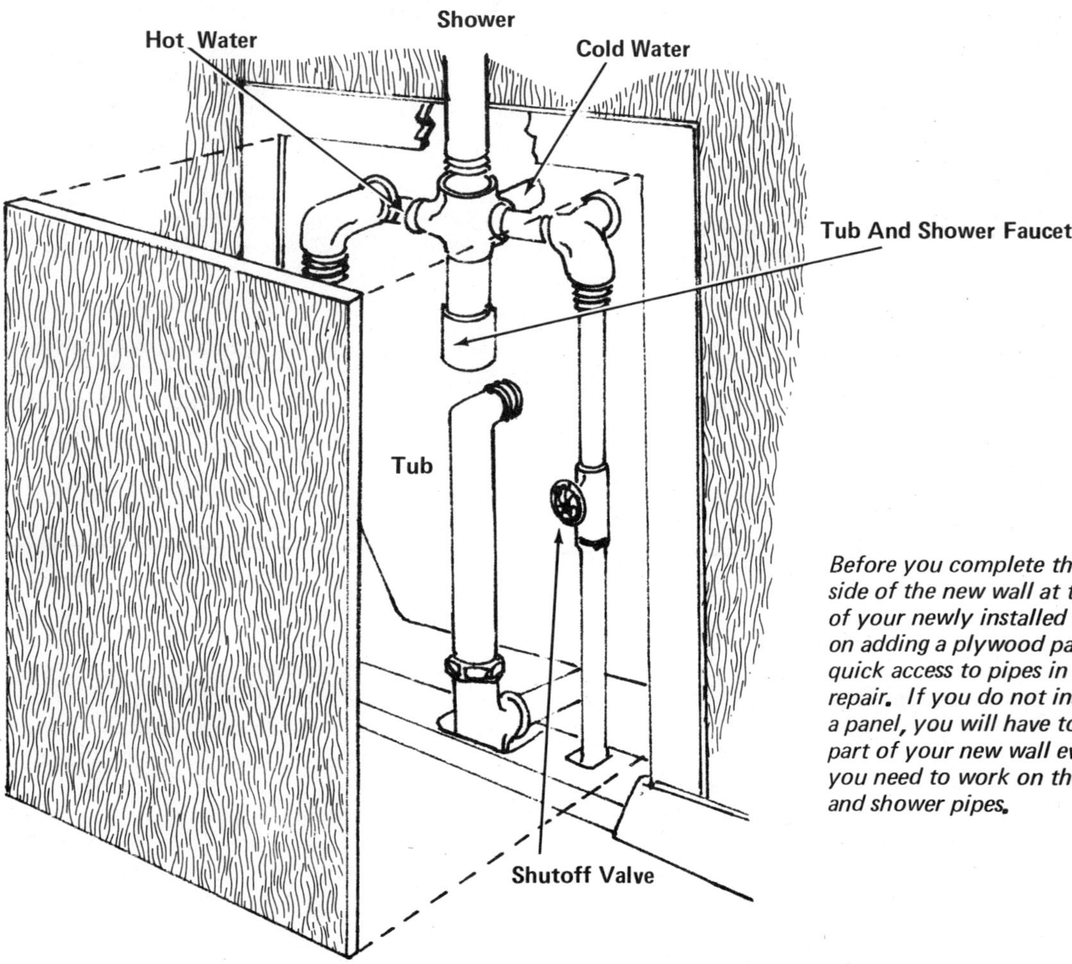

Before you complete the back side of the new wall at the head of your newly installed tub, plan on adding a plywood panel for quick access to pipes in need of repair. If you do not install such a panel, you will have to rip out part of your new wall every time you need to work on the tub and shower pipes.

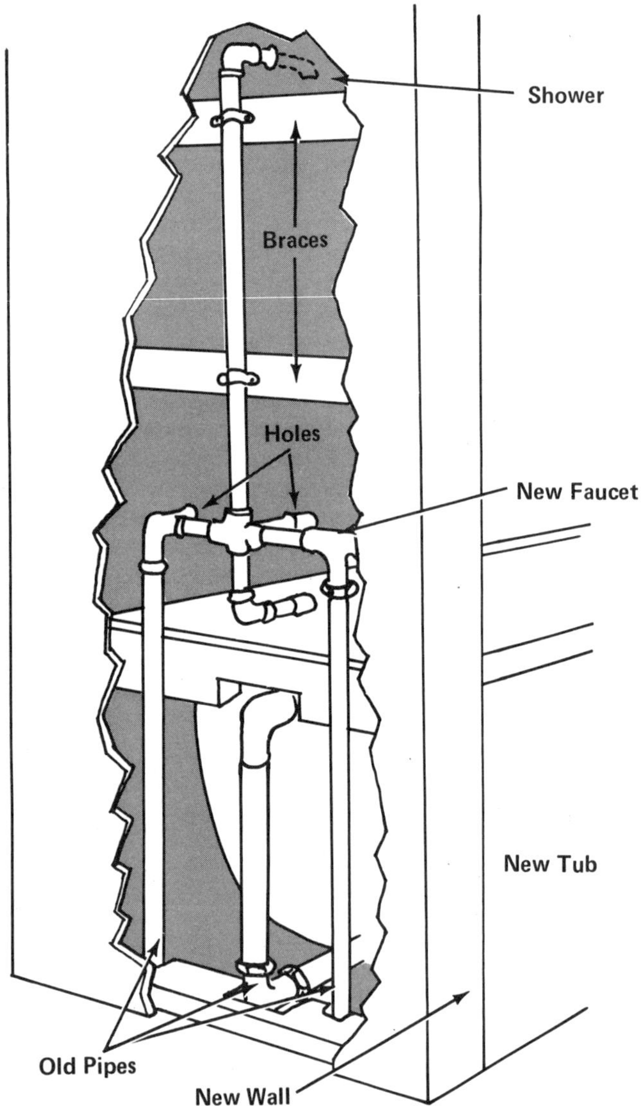

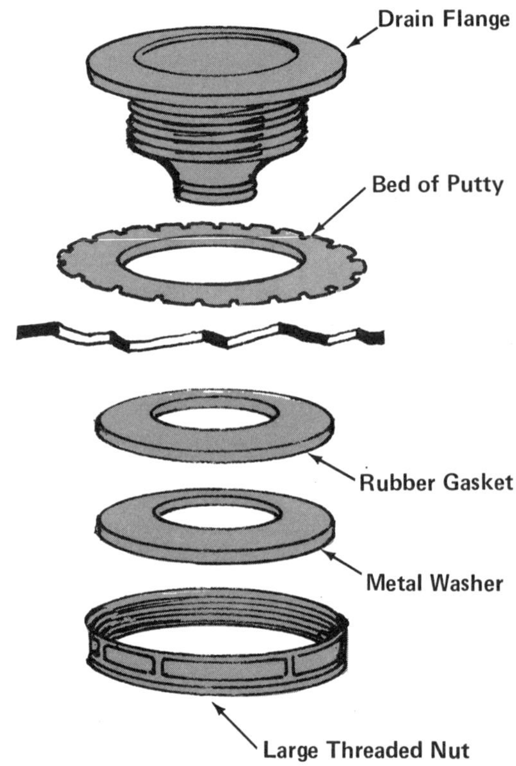

You can use the existing supply and drain pipes when you install a new tub/shower unit (left) as a replacement for an old-fashioned leg-supported bathtub, but you must provide at least one cross brace and U clip in the framing to steady the shower riser. One type of sink stopper (above) is connected directly to the strainer; it employs a screw nut to hold the stopper unit in the sink. When you install a replacement stopper, you must tighten the nut until the unit is waterproof.

fail to block off the outlets, all the pressure you create will be dissipated long before it can get to the clog.

In most cases, you can use a plunger on a kitchen sink without worrying about other outlets; but if you have a double sink, you must block the drain on the other side. In addition, there may be yet another drain outlet connected to the same line that must be blocked as well; you will find that out when the water starts backing up in the unsuspected outlet. With bathtubs and basins, you must close off the overflow vents. In homes that have two bathrooms back to back in adjacent rooms, there is a good chance that both are connected to the same drain. In such cases, you must block the other basin at both the drain and the overflow vent. Shower facilities seldom have overflow vents, but laundry tubs may have two or as many as three such vents. You must cover them all for your plunger to work.

Once you get all the vents covered, place the cup of the plumber's friend over the drain with the handle sticking straight up. Start a rhythmic stroking in a slow and steady motion, and you will soon be able to feel the water move in and out of the drain. It is this back-and-forth water pressure that can eventually build up enough force to dislodge whatever is blocking the drain. Continue to pump the plunger for about 15 to 20 strokes, and then lift it out. The water should go rushing out, but if it fails to move, try the same routine again two or three times before going on to another drain-clearing method.

OTHER DRAIN-CLEARING METHODS

If the plunger fails to clear the drain, at least you have not lost anything. In fact, you probably got

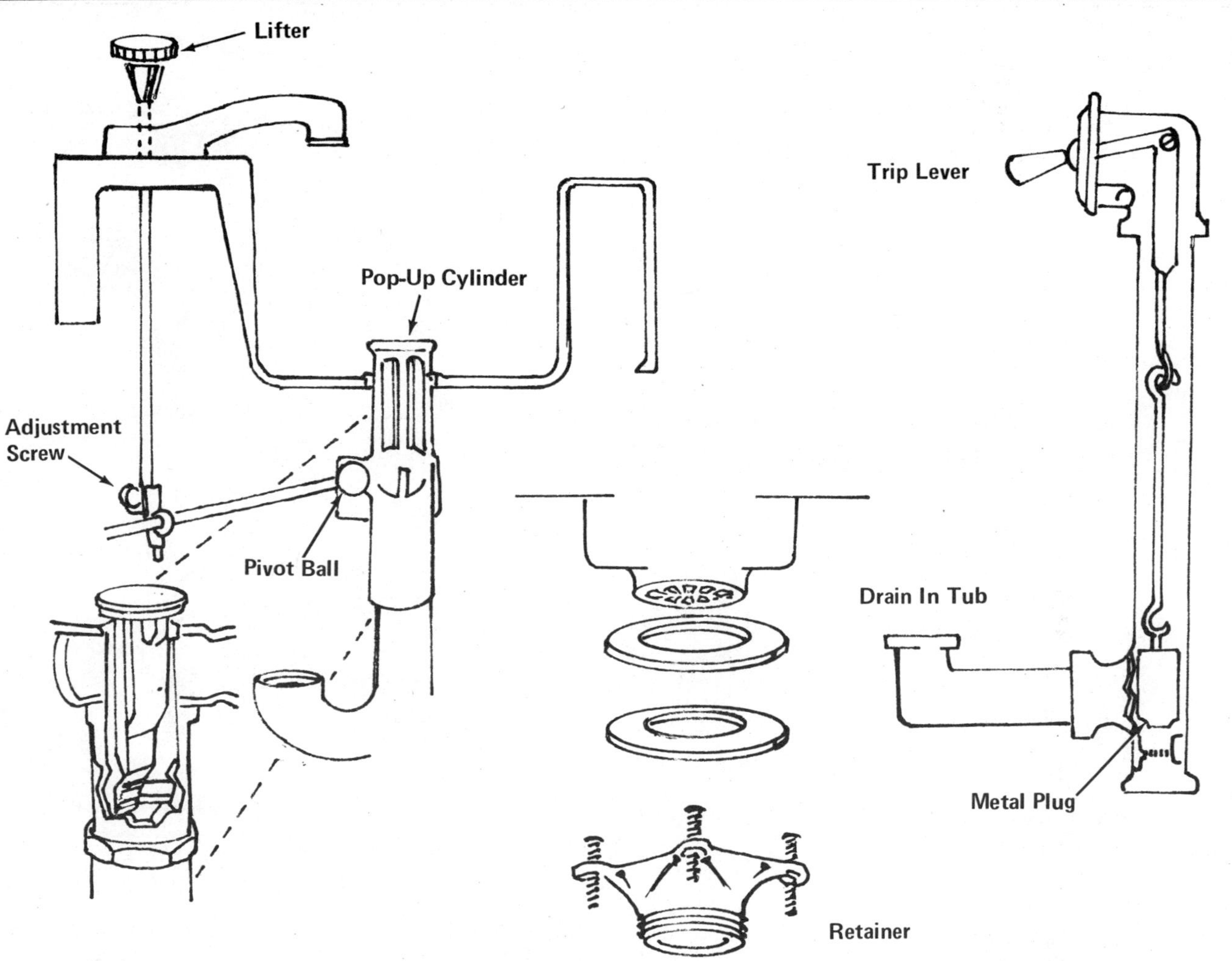

The most common lavatory basin stopper is the pop-up hollow cylinder (left). The top of the cylinder is closed to keep the water in the basin when the cylinder is down. The retainer-type sink stopper (middle) is connected directly to the strainer. When replacing such a stopper, you must tighten the retainer screws against the friction ring until the unit is watertight. The trip lever system (right) closes the drain by lowering a metal plug into such a position that it blocks the drain opening. The open lever plate accommodates any overflow.

some good exercise and worked off some of the frustration caused by the drain being clogged in the first place.

Now, however, it is time to go on to other weapons in the arsenal against drain clogs. Certainly the favorite of most people is the chemical drain opener — in either dry or liquid form. Many people, in fact, use chemicals as a preventative measure, applying them regularly before the drains actually become clogged. Such routine applications will probably prevent any clog from forming in your drains.

Chemicals must be used with caution, however. Only certain types, for example, are suitable for use in a kitchen garbage disposer unit, and all would be harmful were they to make direct contact with your skin. Follow the manufacturer's directions on the label, though, and you need not be afraid; but be especially careful not to make contact with the chemically treated water should the drain opener fail to work. You can try other tools and techniques to get the drain open, but you must be very cautious once you add the chemicals.

Should the plunger and the chemicals fail and you have access to a plumber's snake, by all means use it. The snake, whose proper name is the plumber's auger, is a very flexible metal coil that you can feed down into the drain until it reaches the clog. The snake is so flexible that it can negotiate all sorts of turns and bends in the pipe, and when it finally reaches the clog, the snake has a head that can work its way through to loosen whatever it is that is blocking the drain.

Although the snake can do a great deal of maneuvering, you can save both it and yourself considerable work if you remove the trap first. Some

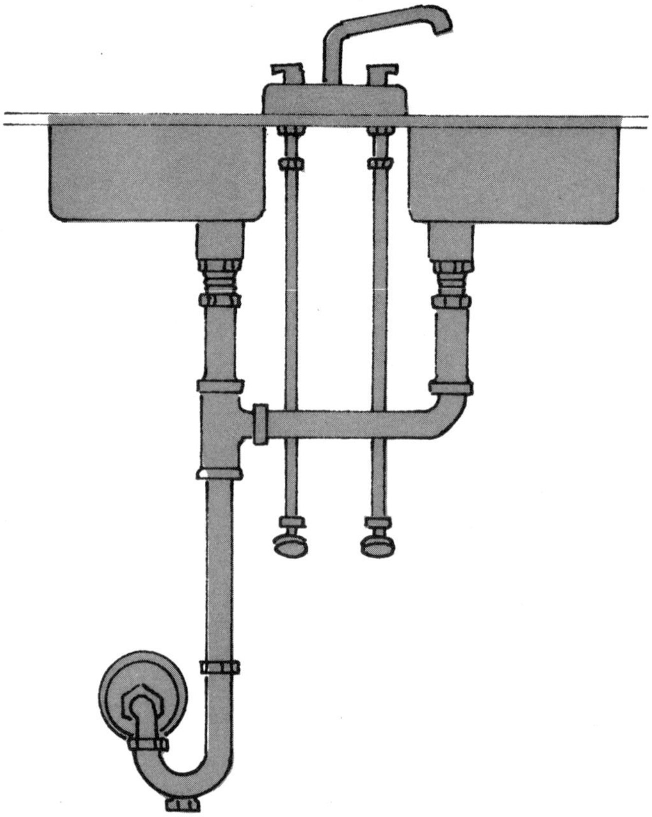

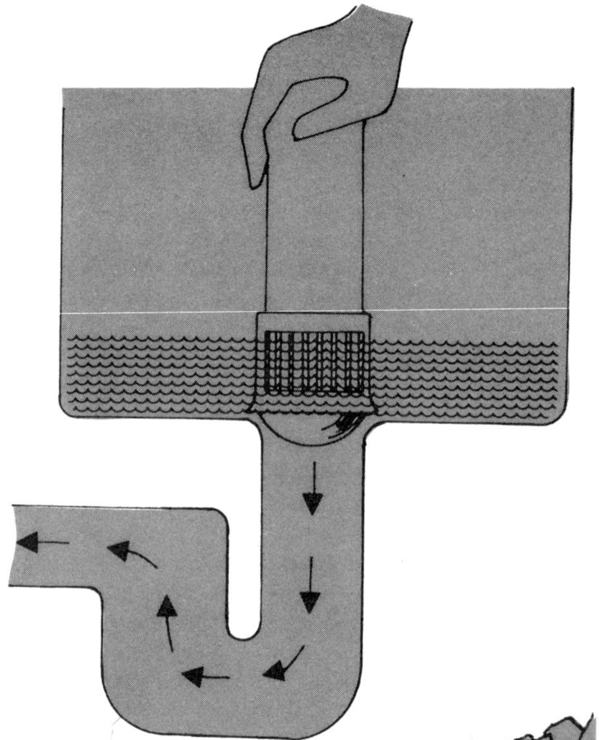

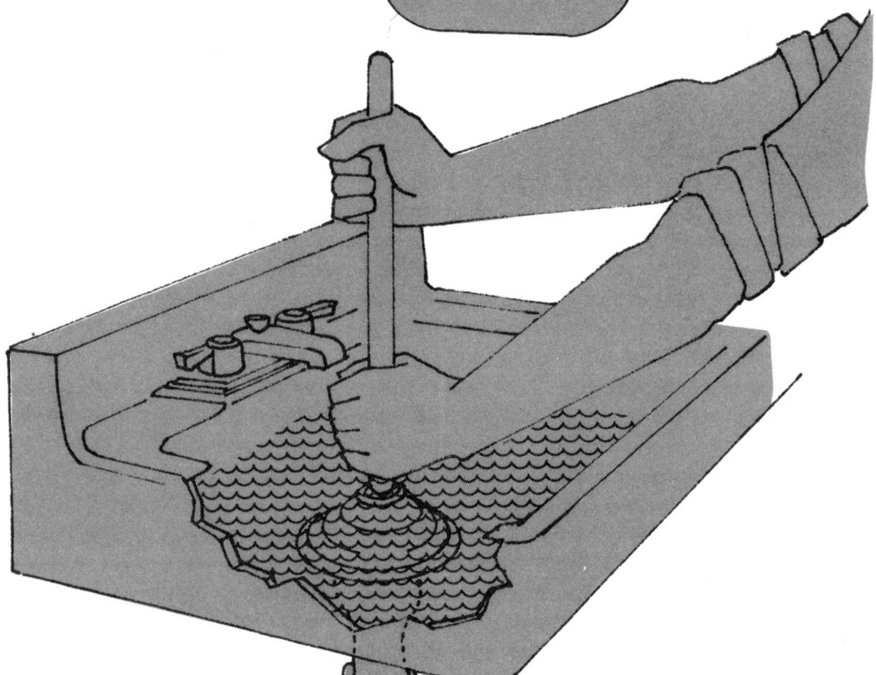

If you have a double sink in your kitchen (above) that gets clogged, then you must block off the drain on one side while you work with your plunger on the other side. Place the cup (right) over the drain with the handle sticking straight up and start pushing and pulling in a slow but steady up-and-down motion. You will soon be able to feel the water moving in and out, hopefully dislodging whatever is blocking your drain. The newest home remedy for a clogged drain is the aerosol can (right, above) that sends a blast of propellant to force the water in the pipe down against the blockage.

traps have a cleanout plug that you can remove instead of the entire trap, but even if you must remove the trap itself, you will find it an easy task. Just turn two slip nuts counterclockwise and slide the nuts along the pipe away from the trap. You will find a rubber washer under each nut; be sure to keep an eye on those washers.

Of course, all the water in the sink will come rushing out when you remove the trap; therefore, have a bucket ready to catch it. And while it may sound too stupid to mention, be sure not to do what all too many people do in this situation — they empty the bucket back into the sink! Once you remove the trap, you may find that you do not even need the plumber's snake because the trap itself may be the location of the clog. Just take a piece of wire and clean out the guck.

Even if the trap is not the site of the clog, however, by removing it you give yourself a straighter shot with the snake into the pipes. Feed the snake in until it stops. It is unlikely you have reached the clog at this point but rather a turn in the pipe. If the snake has a handle, push it up to within a few inches of the opening; then turn the thumb screw tight against the snake to anchor the handle, and turn the handle back and forth as you push and pull the snake in and out. Eventually, the head of the snake will maneuver past the turn. When it does, loosen the handle and slide it out of the way, and keep feeding the snake into the pipe.

You may hit two or three turns on your way to the clog, but be patient. You will finally come to the blockage, and when you do it will feel different than the turns. Again, fasten down the handle and work the snake back and forth as you turn it. The snake head is designed to grab some of the blockage, pull it back out, and push the rest of the clog on through. Keep working the snake to loosen all of the blockage, and then — just to be safe — run the snake a few additional feet into the pipe to spread the debris and move it along.

As you pull the snake out, have your bucket ready under the pipe to catch the bits of blockage that the snake may bring with it. Of course, the snake itself will be wet and filthy, and one of the last things you want to happen is for the snake to spring against and damage your furnishings. Therefore, have a corrugated cardboard box handy, and feed the snake into it as the long tool emerges from the pipe.

Once the snake has done its job, replace the trap, apply pipe joint compound to the threads of the slip nuts, and check the joints for leaks with the first water that you run through. Once you are satisfied that the joints are watertight, turn the hot water on full force for several minutes to flush out all of the loosened debris that once blocked your drain.

There is a new product on the market designed for those people who do not want to use a snake and who have not had good results with a plunger or with chemicals. The product (the first of its type being marketed under the name "Drain Power") is an aerosol canister that sends a blast of propellant down the pipe to blow out the blockage. You merely place the domed head of the canister in the drain and push downward.

Obviously, the aerosol unclogger works on essentially the same principle as the plumber's friend — pressure generated against the clog. This pressure, of course, will seek the path of least resistance. Therefore, if you are planning to use an aerosol to clear your drains, you must block any overflow vent or other outlets on the same drain because the force of the propellant will try to escape the easiest possible way.

The aerosol type of drain opener does not work on clogged toilets, and it cannot be used on a sink or tub that has a nonremovable strainer. The propellant itself is harmless to you, but you must be careful in using it where you have already applied a drain-clearing chemical. To avoid getting splashed with corrosive, make sure that the chemicals are completely flushed from the drain before you use the aerosol.

Here is the basic procedure for using an aerosol drain cleaner:

1. Remove any visible debris from around the drain opening.
2. Locate all other outlets and block them off.
3. Remove any built-in stoppers or strainers. Some strainers are held in place by a bolt that can be removed easily with a screwdriver. Others, held in by spring tension, can be snapped out.
4. If there is no water standing in the sink, put the stopper in and fill the sink or tub to a level of about two inches. If the blockage is such that the water cannot run out quickly, you need not close the drain.
5. With all vents closed, replace the stopper with the domed end of the cannister.
6. Push down sharply on the end of the can for one second only — no more.
7. Remove the can from the drain slowly. If the blast did what it should, the water in the sink will swirl on out. You may see a white mist rise up from the drain opening after unclogging occurs; the mist is just water vapor created by the aerosol.
8. Run the hot water for about a minute to clear out the loosened debris.

There is one type of drain clog that will not respond satisfactorily to any of these techniques: plunger, snake, chemicals, or aerosol. That is when the main drain or a floor drain in the basement gets stopped up from tree roots that have grown in at the joints. You can buy certain chemicals that eat away at the roots, but probably the most effective solution is an electric rooter which goes into the pipe and cuts away roots from the pipe walls as it moves along.

You can rent one of these power augers at a tool-rental firm. When you get it home, feed the auger cable into the cleanout plug closest to the blockage. Soon you will feel the cable strain as the cutting head encounters the roots. Keep feeding the cable slowly until you feel a breakthrough; then go over the area once again. Remove the cable slowly and run water from a garden hose to wash away the root cuttings. Before you return the power auger to the rental firm, replace the cleanout plug and flush a toilet several times. When you are convinced that the drain is clear of tree roots, clean the cable and return the machine.

Chapter Five

Plumbing Related Appliances

While everyone knows that water and electricity do not mix, modern engineering has come up with devices that combine the plumbing and electrical systems to reduce the time and effort required to wash clothes, wash dishes, and eliminate garbage. These are the plumbing-related appliances in your home — clothes washer, dishwasher, and garbage disposer — and you should know how they work, how to fix them, and how to install replacement models if that should become necessary.

Of course, the fact that these appliances are marvels of engineering efficiency does not negate the basic principle that water and electricity can make a lethal combination if you are not extremely careful. Whenever you propose to tackle an appliance maintenance task, fix-it repairs, or new installation, be sure that you first disconnect the appliance from the power source. Either pull the plug or — if that is inconvenient because the unit is built into cabinetry — remove the appropriate fuse or trip the appropriate circuit breaker.

As long as you remember this basic safety principle, you can do a great deal of the servicing that your plumbing-related appliances require without endangering yourself. Naturally, by performing these tasks yourself — and by hooking up new appliances yourself — you can save yourself a great deal of money.

GARBAGE DISPOSERS

Garbage disposers are not usually temperamental appliances — they will eat just about anything you serve them. The most expensive models can even digest bones, corncobs, and seafood shells. Some day, however, you are going to feed your disposer something that it cannot digest — bottle caps, cans, glass, crockery, rags, string, paper, cardboard, rubber, plastic, cigarette filters, etc. — and the disposer will let you know of its indigestion by ceasing its operations.

You can cure disposer jams as soon as you understand exactly how the appliance works. Contrary to what most people think, the blades in a garbage disposer are not sharp and they are not designed to cut up the waste. Instead, the blades (called impellers) pivot on a flywheel, battering the garbage and slinging it against the shredder and cutter surfaces of the unit. The flywheel spins at a rate of over 1700 revolutions per minute, but it will come to a sudden halt if particles of hard matter get lodged between moving parts.

You can clear such jams quickly and easily through one simple procedure. Once you make sure that your garbage disposer is turned off, insert the end of a broom handle, angle it against one of the impeller blades, and pry. The idea is to use the broom handle as a lever to get particles out from where they are lodged under the moving parts. Eventually, you will feel the impeller give — although you may have to repeat the procedure several times — and you can then withdraw the broom handle. Once you clear the blockage, push the disposer's reset button (it trips automatically when a jam occurs), and you are ready to feed more garbage into the hungry unit.

Some garbage disposers have a crank of some sort that fits into a socket in the bottom of the unit and lets you turn the flywheel plate in reverse. Deluxe models often have an automatic reversing mechanism that you operate merely by flipping the switch off and then on again. Whether done manually or automatically, however, the purpose of the reverse action is the same: to loosen the clog.

Although you should be able to clear any clog in your disposer without calling for professional assistance, there is really no reason to get into such a predicament in the first place. Be kind to your disposer; refrain from feeding it objects that are bad for its continued good health. If you are ever in doubt as to whether a particular substance should go into the disposer, play it safe until you know for sure and

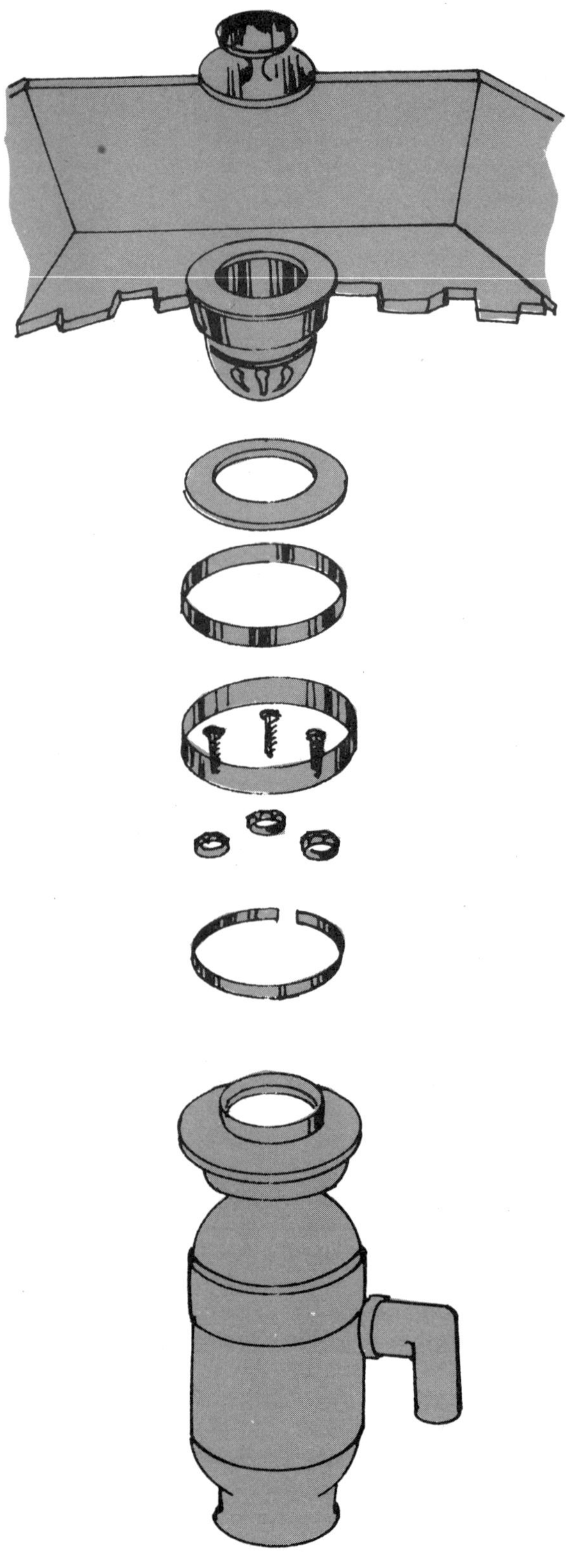

Installation of a garbage disposer, either as a new addition to your kitchen or as a replacement appliance, certainly constitutes a feasible do-it-yourself project. Detailed instructions are included with the unit.

place the questionable material in your garbage pail instead. You may occasionally get away with depositing some of the things on the forbidden list, but eventually you will damage the disposer or clog the drain — or both.

Always use cold water when you run your garbage disposer. Many people turn on the hot water tap, thinking that the hotter the water the better the grease will drain away. Cold water, however, lets the grease congeal into tiny solid bits that can then be washed down the drain. Hot water only melts the grease, which then coats both the garbage disposer and the drainage pipes.

As grease builds up on the walls of the disposer, food particles collect there and cause unpleasant odors. Be careful, though, about using drain cleaners in your disposer unit. Never put regular drain cleaning chemicals down your kitchen disposer. There are drain cleaners that are made specifically for disposers, and you should use only those types that are labeled as safe for use in garbage disposers. Such cleaners contain a petroleum distillate that can degrease, unclog, clean, and deodorize the unit.

Unless your garbage unit really needs a chemical cleaning, in fact, try this deodorizing technique. Fill the disposer about three quarters full of ice cubes, turn the unit on, and ignore the loud noises that ensue. Ice cubes can help to scrape off the food residue that is clinging to the disposer's walls. When all the ice is chewed up, flush the unit with cold water. Finally, drop in half of a lemon and let the disposer chew it up; you should discover that the nice lemon smell totally replaces the former foul odors.

Preventive maintenance is about the limit of what the do-it-yourselfer can do in terms of garbage disposer repairs (although repair kits are available for some units in the event that an impeller blade comes loose). Most units are sealed and cannot be repaired easily. If you decide that your garbage disposer needs repair, however, you are probably wisest to call the authorized repair station for the brand you own. Stop and consider this point, though, before you make the call: You can probably put in an entirely new disposer unit for very little more than you would pay a professional repairman to fix your old one.

Installation of a new garbage disposer certainly constitutes a feasible do-it-yourself project, but there is one fairly unpleasant chore you must perform before you can put in a new or a replacement unit. You must thoroughly clean out the drain lines with a plumber's snake to get rid of the small particles of waste that would eventually build up into a clog when all the grease starts rolling through. Once you get that done, you can put in the new unit. You will get detailed installation instructions with your disposer, but here are the typical steps required to add a garbage disposer to your kitchen sink:

1. Remove the drain pipe under the sink.

Anyone can replace a worn-out garbage disposer. Only a few of the most common household tools are required. You can do the entire job at less cost than it would take to repair your old unit.

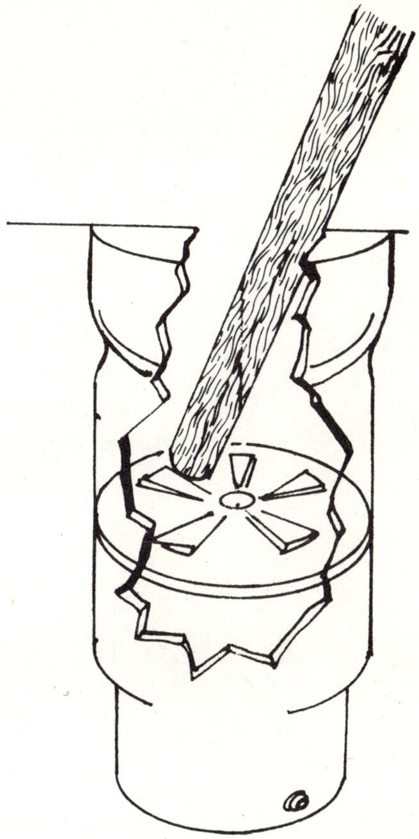

If your garbage disposer ever stalls, try sticking the end of a broom handle in it with the unit off. Use the handle as a lever and pry against the impeller blades to free them from lodged particles.

Garbage Disposer Troubleshooting Chart

PROBLEMS	CAUSES	REPAIRS
Garbage disposer jams	1. Nondisposable substance deposited in disposer.	1. Insert broom handle and pry against jammed impellers. 2. Turn flywheel with crank on manual-reverse models. 3. Flip switch off and then on again with automatic-reverse models.
Unpleasant odors coming from disposer	1. Grease buildup traps food particles.	1. Always use cold water to flush garbage disposer in order to prevent grease buildup. 2. Apply drain cleaner, but only the kind specifically made for garbage disposers. 3. Fill disposer about 3/4 full of ice cubes and turn unit on. Drop in half of a lemon after cubes are chewed up and unit flushed with water. Let unit chew up lemon as well.

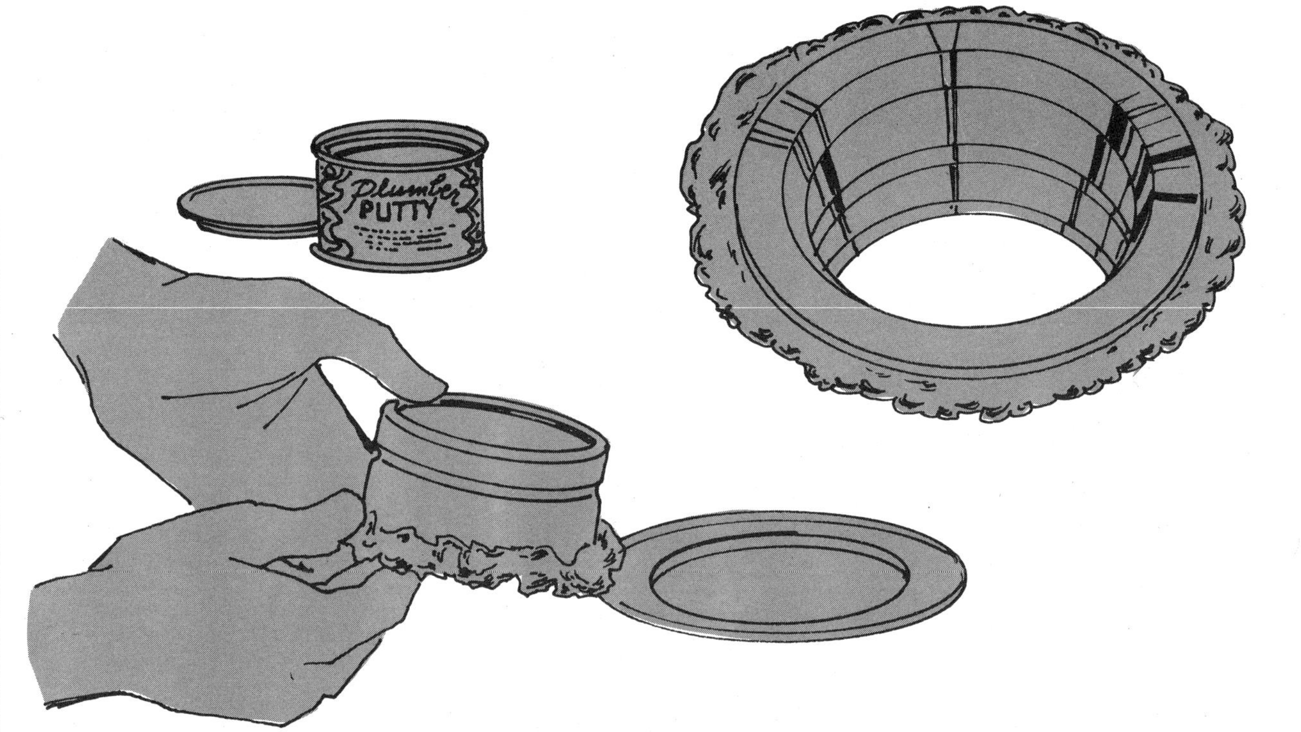

Place a bed of plumber's putty around the opening in the sink and around the underside of the disposer's sink flange.

2. Remove the old sink flange, as well as the sealing material and any gaskets from the sink.
3. Place a bed of plumber's putty around the opening and around the underside of the sink flange unit.
4. Insert the flange into the opening and press it down into place. Once it is in place, do not rotate the flange; otherwise, you could break the seal.
5. Remove the excess putty from around the flange.
6. Now you are ready to go below the sink. Slip the gasket over the underside of the sink flange, followed by the protector ring with the flat side up. The mounting ring — which has three threaded pins inserted in it — follows the protector ring.
7. While holding these parts in place above the groove on the flange, push the snap ring up along the flange until it snaps into the groove. One word of caution about the snap ring: If you spread it, it can become too loose to hold in place around the groove. You may find it slow to push the snap ring in place, but it will go. Make sure it fits firmly in the groove.
8. The threaded pins have screw slots which must be uniformly tightened against the protector ring to hold it and the gasket snugly against the bottom of the flange.
9. Now, lift the disposer unit and put it in place. Match the holes in the disposer with the threaded pins, but before tightening the nuts that hold the unit in place, rotate the disposer so that the outlet pipe faces in the right direction. Tighten the nuts.
10. Now you are ready to hook up the drain. Use slip nuts and remember that the parts of the trap can be maneuvered around so that they will fit together. If this is a replacement unit, you will have no problem. If it is a new installation, you may have to buy and install some extra pipe sections. A common trap for both the disposer and the other side of the sink is acceptable.
11. Test the installation and the unit for leaks.
12. Before you plug in the unit, reach inside to see whether you dropped any tools in there. Now you can start throwing the garbage down the sink.

CLOTHES WASHERS

An automatic clothes washer probably constitutes the greatest labor-saving plumbing-related appliance in the home. A complete discussion of how one works and how to fix one when it stops working would be beyond the scope of this book, but you should have an idea of the plumbing problems that an automatic clothes washer can display. Moreover, if you know how to put in a new unit, you can save yourself a good deal of money on installation charges.

Most of the plumbing problems that occur in an automatic clothes washer have something to do with the water-inlet valves. These valves are operated by solenoids which raise plungers to allow water to flow into the machine and which lower plungers when the water flow is supposed to be shut off. The number of solenoids depends upon the number of water-temperature selections on the clothes washer. There may be one (hot only), two (hot and cold), or three (hot, cold, and warm — a mixture of hot and cold) solenoids. If one or more solenoids malfunction, you find that water does not enter your machine.

There are other causes for water failing to flow into your clothes washer. In addition to the solenoid, the water-inlet valve itself may be defective. If it is, you must replace the valve before the water will enter the machine. In cases where both the solenoid and the water-inlet valve are in good working order, the problem may simply be that a water inlet hose is crimped or that the faucet to which the hose is connected may have been turned off for one reason or another. You can, of course, repair such difficulties merely by straightening the crimped hose or opening the closed valve.

A somewhat more involved procedure is required, however, if none of the above solutions gets the water to flow. Disconnect the water-inlet hoses at the faucets and at the machine's water-inlet valves, and check the filter screens at both locations. It is quite possible that sand or rust or sediment could be clogging the screens and preventing the water from flowing into the machine. Clean the filter screens with an old toothbrush, replace them, and try the washer once again. If the water still does not enter, install new filter screens. Frequently, a thin transparent film of algae can remain in the screens even after thorough cleaning. New filter screens should, however, restore your machine to good working order.

As serious a problem as no water entering your machine may be, it is nothing compared to when the water does not stop and your machine overflows. The first thing to do, of course, is to unplug the washer. If the water stops, then you know the problem lies in an electrical component — either the solenoids on the inlet valves or the machine's water-level switch is malfunctioning and must be replaced.

If the water continues to flow even after you pull the plug, then the problem is in the water-inlet valve itself. These valves depend upon balanced water pressure on both sides of a diaphragm. The balance can be thrown off by some sand or rust in the water that somehow gets past the filter screens at the hose terminals. You can disassemble the water-inlet valves to clean them, but it is a good idea to

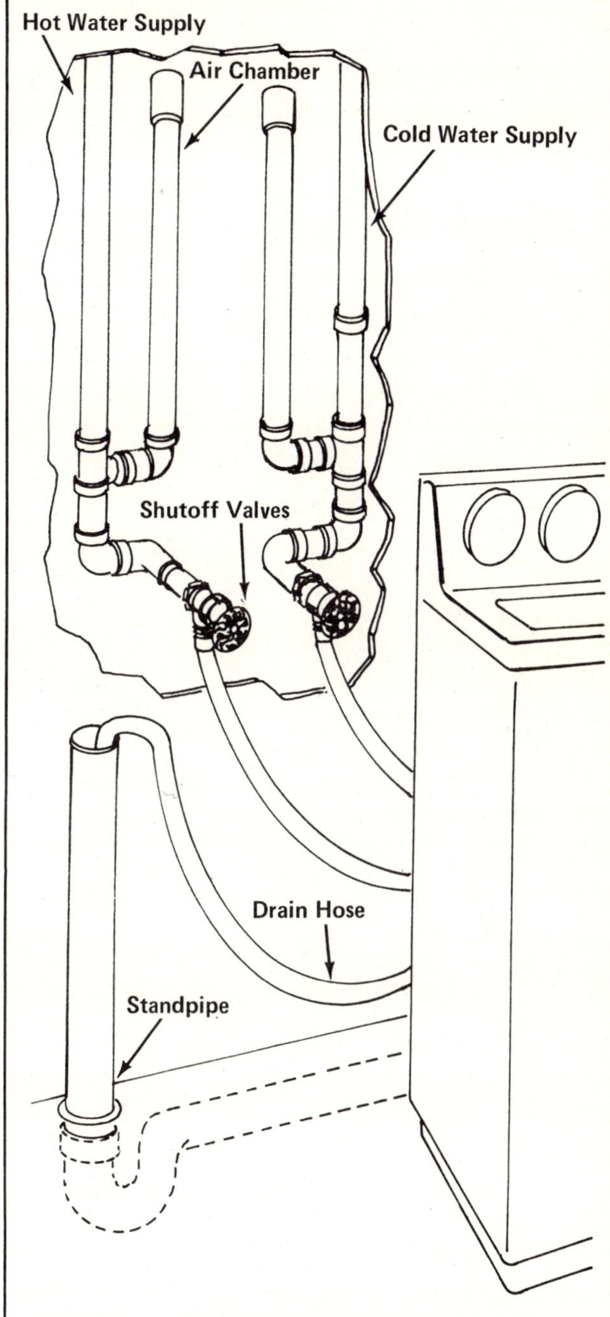

If you install a clothes washer yourself, be sure that the standpipe is higher than the highest level that the water in the machine reaches. Threaded cutoffs on the supply pipes make the job simple; just attach threaded hoses.

Clothes Washer Troubleshooting Chart

PROBLEMS	CAUSES	REPAIRS
Washer does not fill with water	1. Water cutoff valves on pipes closed.	1. Open valves.
	2. Water-inlet hoses kinked or knotted.	2. Straighten out hoses.
	3. Clogged water-inlet screens.	3. Screens are located in hose connections at water faucets and at washer. Remove screens and clean out sediment.
	4. Defective water-inlet valve solenoid.	4. Repair or replace it.
	5. Defective water-inlet valve.	5. Disassemble water-inlet valve and inspect parts for damage. Replace a bad part with a new one, if possible. If not, replace the entire valve.
Washer overflows	1. Faulty solenoids on water-inlet valves or faulty water-level switch.	1. Pull plug. If water stops, then problem is in an electrical component. Replace defective component or call repairman.
	2. Clogged water-inlet valve.	2. Disassemble valve and clean it to restore balanced water pressure on valve. Turn off water cutoff when machine not in use to relieve damaging pressure against valve.

prevent valve damage before you are faced with an emergency situation. Use the machine's cutoff valves whenever you are not washing clothes in order to relieve the water pressure against the water-inlet valves. Then, at least, you will be facing merely a plumbing problem — not a disastrous flood — if a water-inlet valve should malfunction.

If you plan to install a new machine, you will need to have both hot and cold water supply pipes handy. Threaded shutoff valves will make your washer installation task much easier because you can then simply attach threaded hoses to bring the water to the washer. Generally, there are hoses already attached to the washer, and all you have to do is screw the hoses to the cutoffs. A rubber hose washer insures a tight connection.

You should also have a drain nearby to which you can connect a standpipe. The standpipe should be higher than the highest level that the water will reach in the machine. You need not actually attach the washer's drain hose to the standpipe; just make sure that the hose will stay in place. Push it down into the standpipe about six inches, and then surround it with wedges cut from a section of an old automobile tire. In fact, you want there to be a source of air around the drain hose in the standpipe as a safety precaution; otherwise, there exists the possibility of a cross connection between the washer's waste water and the plumbing supply lines.

If you have to install entirely new plumbing when you put in your new clothes washer, be sure to provide air chambers. When you are just attaching hoses to cutoffs, there should be no problem; most likely, the pipes are already equipped with air chambers. But if you have to bring the water supply to the unit via new plumbing, it is essential that you install the chambers as well. An automatic washer turns the water off and on so often that if there were no air chambers, the water hammer would be unbearable.

Here is a hint that has nothing to do with plumbing, but which can give your automatic clothes washer much longer life. When you finally get the unit all hooked up, get out a carpenter's level and make sure that one side of the washer is not higher than the other. Even a small degree of imbalance places an extra strain on the motor, thereby reducing both the efficiency and the longevity of your washer.

An automatic dishwasher is easy to operate, but it can suffer from many plumbing problems, including valve failure.

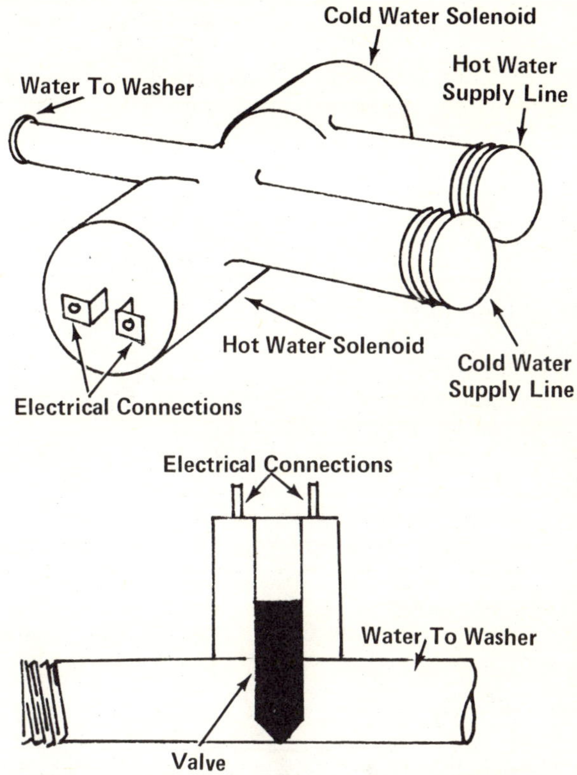

Water-inlet valves on an automatic clothes washer are controlled by solenoids. If the washer has just one temperature setting, the water-inlet valve is merely an on-off device (below); double-solenoid valve (above) allows for hot, cold, and warm temperature settings.

DISHWASHERS

Dishwashers are, at least insofar as their plumbing is concerned, similar to automatic clothes washers. Most of the plumbing problems you will encounter probably have something to do with the water-inlet valves. For example, if water does not enter the dishwasher, the filter screen in the valve is probably clogged or else the solenoid that operates the valve is malfunctioning. Similarly, if the water fails to shut off, the valve itself or the solenoid that controls it is probably at fault. Very often you can remove the inlet valve or the inlet-valve screen and clean off any residue. Solenoid repair or replacement requires a bit more technical expertise.

Your dishwasher may seem to operate properly, however, and yet your dishes may come out less clean than you think they should be. In such instances, check both the water pressure and the water temperature — two extremely important factors in dishwasher performance. The water supply must flow at a rate of at least two or two and a half gallons per minute and at a pressure of at least fifteen pounds per square inch. The ideal water

If your dishes do not come out clean, the problem may be that the water is not hot enough. 150 degrees is ideal.

temperature is 150 degrees, but anywhere within the 140 to 160 degree range is acceptable. Water that is not at least 140 degrees cannot dissolve grease or detergent. If you run your fingers around your machine and pick up a greasy black film, then you know that the water temperature is too low.

Leaks from an automatic dishwasher usually do not occur at the plumbing connections. They are more often the product of door gaskets that have been pulled loose or that are in some other way defective. If, however, you do discover a leak in a hose connection, be sure to pull the plug, trip the circuit breaker switch, or remove the fuse before you reach in under your dishwasher to tighten the loose hose. Were you to touch a live wire while working on a faulty plumbing connection you would put an early end to your do-it-yourself career.

Installing a new dishwasher is easier than putting in some other fixtures because you must only hook up one supply line, the hot one. If the hot water supply cutoff is threaded, then it is merely a question of attaching the hose from the dishwasher to the faucet-type cutoff.

Most of the connections that will concern you involve the dishwasher's drainage rather than its supply. If the unit drains directly into your home's drainage lines, then it needs a trap like any other plumbing fixture. Install the necessary pipes and trap according to whatever procedure your plumbing situation dictates.

If the dishwasher drains through the same pipe that handles the flow from your kitchen sink, then you must install a dishwasher branch tailpiece to replace the regular tailpiece of the kitchen sink drain. The dishwasher branch tailpiece has male threads, and you attach the drain hose from the dishwasher directly to the tailpiece. The connection directs the dishwasher's waste water out through the sink trap, and a rubber washer inside the connection insures a watertight fit. Since the dishwasher has a pump to remove the water from the unit, you need not worry about whether the drain connection is above or below the water level in the dishwasher. All the water will be expelled in either case.

Many people have their dishwasher drainage hose directed into their garbage disposer. There is nothing wrong with such a connection so long as you make sure that the disposer is kept free of clogs at

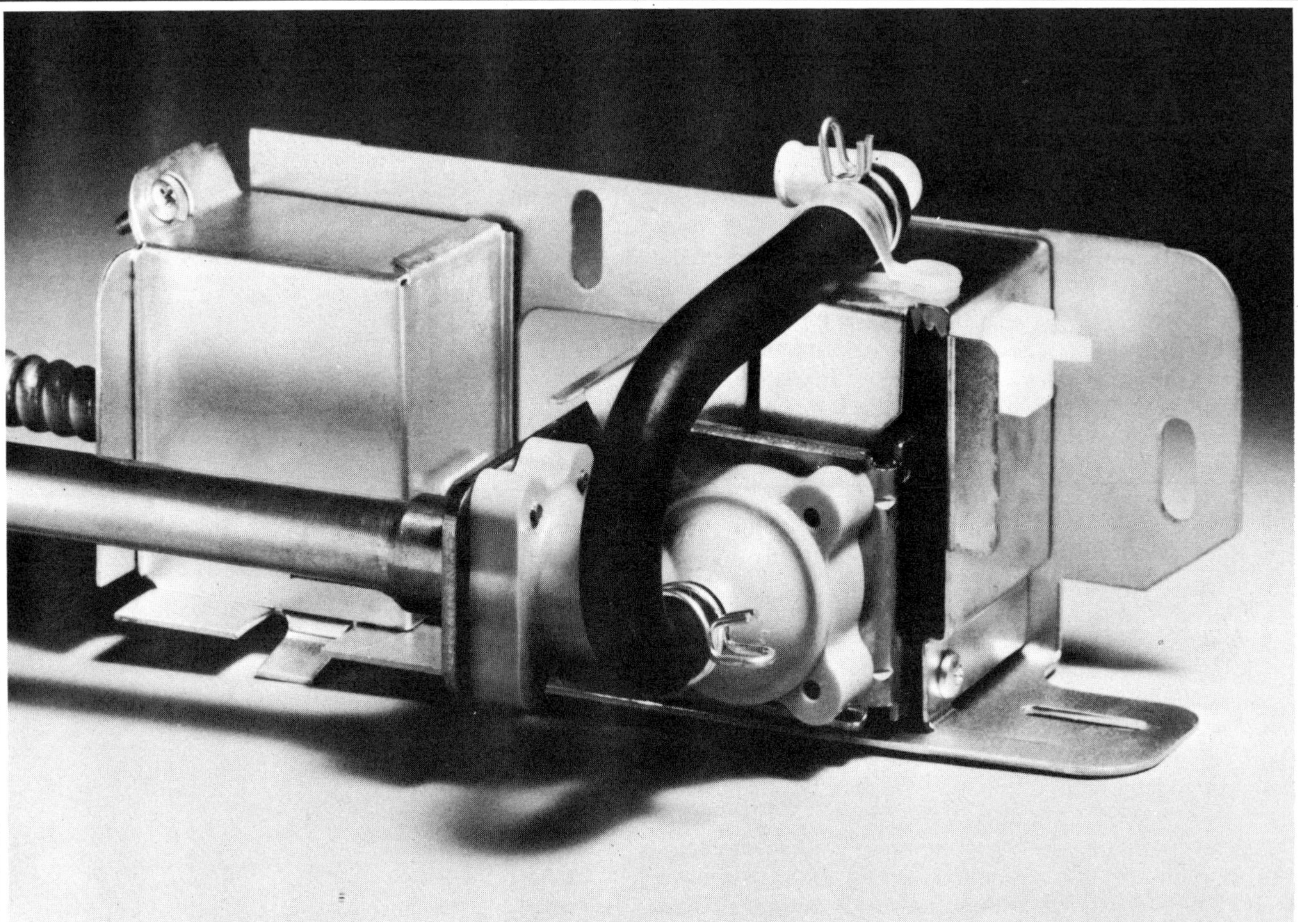

For the most part, malfunctions in the motor assembly of a dishwasher should be repaired only by a professional technician.

Dishwasher Troubleshooting Chart

PROBLEMS	CAUSES	REPAIRS
Dishwasher fails to fill	1. Water valve partially or fully closed.	1. Check water valve on line leading to dishwasher and open fully.
	2. Water pressure too low.	2. Water pressure must be at least 15 pounds per square inch; call your local water company.
	3. Clogged water inlet screen or damaged water valve.	3. Most water valves are equipped with screens to trap deposits. Disassemble and clean screen. If parts are heavily calcified, replace entire water-inlet valve assembly.
	4. Defective water valve solenoid.	4. Turn timer dial to "Fill" and check solenoid. If voltage exists at solenoid, but water valve is not opening though free, solenoid is defective. Repair or replace it.
Dishwasher does not stop running water	1. Water-inlet valve stuck open.	1. Disassemble and clean it.
	2. Clogged water-inlet valve bleed hole.	2. Disassemble valve and clean out bleed hole.
	3. Defective solenoid.	3. Check solenoid, and repair or replace it if necessary.
Dishes come out greasy	1. Insufficient water pressure.	1. Water supply must flow at 2 to $2^1/_2$ gallons per minute at a pressure of at least 15 pounds per square inch. Call water company.
	2. Insufficient water temperature.	2. Ideal water temperature is 150 degrees, but anywhere in the 140 to 160 range is acceptable.
Dishwasher leaks	1. Defective door gasket.	1. Tighten gasket that has been pulled loose, or replace gasket that is defective.
	2. Loose hose connection.	2. Disconnect electricity and tighten the loose hose.

all times to avoid a backup into the dishwasher. In order to be hooked up to the dishwasher, the disposer must have a short inlet pipe on its side; most disposers do. To make the connection, you must remove the knockout plug from inside the disposer. Just angle a screwdriver against the knockout plug and tap the screwdriver with a hammer until the plug comes loose. From there on, follow the instructions that the manufacturer provides for installing the new dishwasher.

Naturally, whenever you add, replace, or repair any plumbing-related appliance, you must make sure that the work you plan to do falls within the provisions of your local plumbing code. Moreover, remember that these appliances are like any other plumbing fixtures in that they must be properly trapped. Finally, keep in mind that these appliances — unlike other plumbing fixtures — are also tied into another power source, electricity. Never even think about repairing or installing any of these devices without first disconnecting it from its electricity supply lines. Always pull the plug, trip the circuit breaker switch, or remove the fuse before you tackle any plumbing-related appliance task.

Chapter Six

Plumbing Products

When most people consider buying a car, a washing machine, a toaster, or a set of tools, they are naturally inclined to select brand-name merchandise. Yet, when it comes to purchasing plumbing products, few people know the difference among brand names and fewer still know whether they should even pay any attention to brand names. There is, unfortunately, no conclusive answer; it depends on what type of plumbing product you plan to purchase.

In some cases, your selection of a brand may already have been made for you by the hardware or plumbing supply dealer. No store can carry every brand of faucet replacement or drain cleaner because there would not be room for anything else. Every dealer must select one or two brands that are competitively priced and that will satisfy his customers. If you can find a hardware or plumbing supply dealer that you trust, you might as well go ahead and accept his selections.

Moreover, with some plumbing materials, brand names mean nothing. For example, pipe and tubing of the same metals in the same sizes and gauges are all about the same. Although makers of plastic pipe often try to make a great deal out of their brand names, the type of plastic involved is much more important than who makes it. The same is true regarding the proper solvent for joining the plastic pipe. While solvents do carry brand-name labels, the important thing is not the brand but rather that the solvent is the right one for use with the particular type of plastic. One brand of plastic pipe or plastic pipe solvent has not proved its superiority over the competition.

Another plumbing product where brand names are of little concern is faucet washers. The key to buying faucet washers is to get the right type and size. Once you match the right washer with the proper installation, your faucet repairs are sure to be successful.

Despite all these examples to the contrary, however, there are some plumbing products where brand names do make a difference. The following recommendations represent plumbing products that are of good quality, easy for the average do-it-yourselfer to install, and which are fairly well distributed throughout the nation. If a particular brand is not mentioned, though, it does not necessarily mean that it is inferior. This list is meant to be representative, not all-inclusive. There are, undoubtedly, other good products in each category. On the other hand, you should certainly be satisfied with any of the products mentioned here.

FAUCETS

When you need to purchase a new faucet unit, there are a number of brand names from which to choose. What you should look for is a good unit that functions well and that comes with helpful installation instructions. You should also know whether the manufacturer offers repair kits and replacement parts that are widely distributed. Even

When you buy a single-handle faucet for do-it-yourself installation, make sure that repair kits are available.

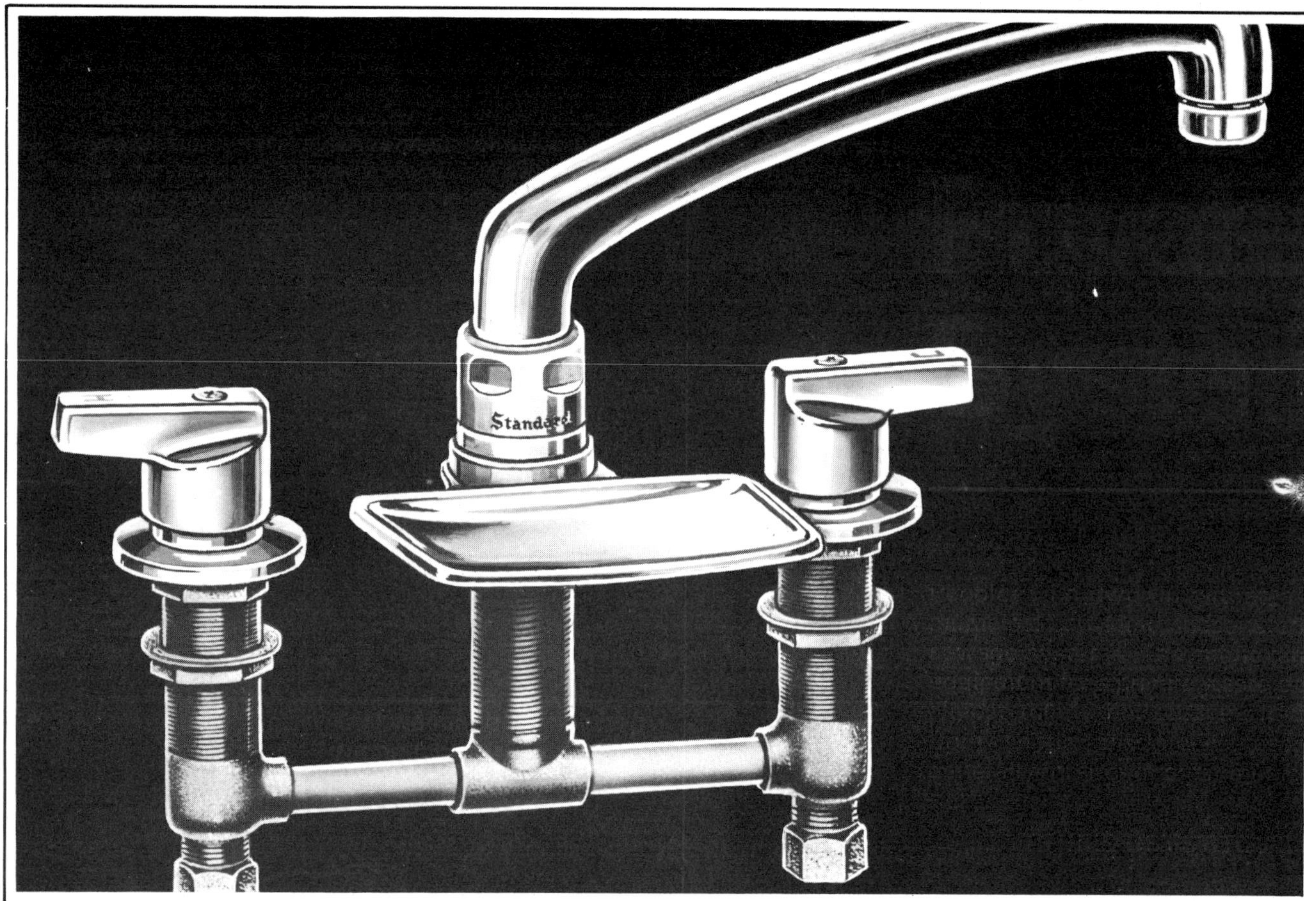

American Standard, a leading name in plumbing products, makes two-handle as well as single-handle faucets.

though many new faucets carry a five-year guarantee, you should make certain that repair kits and/or parts are readily available and that most repairs fall within the do-it-yourself category.

Single-lever faucets for kitchen, basin, and tub and shower units are usually selected on the basis of personal taste in design. Stick with Rockwell, American Standard, Stanadyne, Moen, Peerless, Delta, Union Brass, or Price-Pfister and you should not go wrong. All are priced competitively, engineered well, and designed for easy do-it-yourself installation. It would be difficult — if not impossible — to single out any one of these faucet manufacturers as the best. Some of the companies offer several different quality lines, and since price differences are usually minor, select the best faucet you can afford.

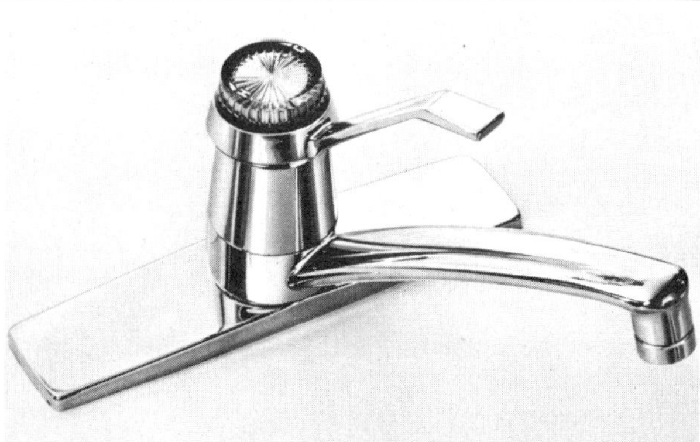

All of the brand-name single-handle replacement faucets are engineered well and priced competitively.

Some of the firms mentioned above also make two-handle faucets as well as single-handle models, but you can add the names Plumb-Rite, Alpha, and Crane to the double-handle list. The two most popular manufacturers are American Standard and Crane, but the others also make good faucets for do-it-yourself installation.

If you are thinking about a two-handle faucet replacement, then consider a Dishmaster unit. Not only is the Dishmaster a top-quality faucet, but it also features a brush attachment with a soap or detergent spray instead of the regular water spray. Dishmaster units fit all standard kitchen sinks, and repair kits are available at the same dealers that sell the units.

If you have a conventional kitchen faucet with a spray attachment, you can easily convert the sprayer head to one that feeds detergent at the push of a button. The replacement, called a Dish Quick, has a large sprayer head and a compartment for the detergent supply. You must refill the Dish Quick sprayer with detergent more often than a Dishmaster unit requires, but if you do not need to replace the kitchen faucets — yet want a sprayer that dispenses suds as well as plain water — the Dish Quick is the one to buy.

CHEMICAL DRAIN CLEANERS

Although there are many chemical drain cleaners on the market, there are few appreciable differences in their ability to unclog drains. Unfortunately, none of them is effective for completely clogged drains, but nearly all of the name-brand chemical drain cleaners are excellent for clearing slow-running drains and for preventive maintenance.

Two brands of liquid cleaner that are available widely are Drano and Liquid-Plumr. Liquid-Plumr does a better job on bathroom clogs where hair is a big problem. As for kitchen blockages, both liquid cleaners get the job done. Mitee, a line found in most hardware stores, is an excellent liquid drain opener and cleaner. Other good brands include Hi-Power and Rooto, both of which are found primarily in hardware stores. All of the crystal type drain cleaners are a little less effective than the liquids, and none of the well-known crystal cleaners works better than ordinary lye.

The newest entry in the drain cleaner field is an aerosol product called Drain Power, available in supermarkets and hardware stores. It is a plastic cannister that shoots a blast of propellant down the drain with enough force to loosen many clogs. Drain Power does not work if the clog is beyond the vent stack, though, because the force of the propellant will seek the path of least resistance and go up the vent pipe rather than strike the clog. The advantage Drain Power offers over the liquid and crystal cleaners is that when it fails, there are no strong chemicals in the drains to endanger your safety.

MECHANICAL DRAIN CLEANERS

Although most of the plungers or plumber's friends are about the same, there is one that does a far superior job on a clogged toilet: the Toilaflex Toilet Plunger. In addition to its conventional rubber cup, the Toilaflex has an extension that is shaped to fit down into the opening of the toilet bowl. The extension centers the suction cup so it cannot move around, making for a tighter fit and thus eliminating the splashback that can occur when you use a standard plunger in a toilet. For sinks and basins — where the drain openings cannot accommodate the tapered tailpiece — the extension can be folded back into the cup. Nevertheless, for sink and basin drain problems, the Toilaflex Toilet Plunger is not as effective as a plain plumber's friend.

There are a number of plumber's snakes or plumber's augers on the market. Usually, retailers carry only one manufacturer's line, but generally there are several types within that line from which to choose. If you have such a choice, pick the heavier auger. Chicago Specialty Company, Cobra Products, Plumb Craft, and General Wire all make snakes that perform as well as you could desire.

Snakes that are housed in their own drums save you from handling the tool itself, and they allow you to feed the snake into the pipe by cranking a handle. General Wire's Spin-Thru, Chicago Specialties' Mighty Mite, and Wrightway's Little Pete and Big Pete are all recommended for most home purposes. If you own an electric drill, you can get a snake attachment for it. The snake is sheathed in plastic tubing so that it cannot harm porcelain sinks or basins. Although the power-driven augers will certainly go through any blockage, the models now on the market are of limited usefulness because they are made only in the shorter lengths.

Another type of drain unclogger utilizes water pressure to blow out the blockage. You know, of course, that if you stick a hose down into a clogged drain, as soon as you turn on the water it will rush down the pipe until it reaches the clog and then rush back and squirt out your end of the pipe. A gadget called a Drain King, however, can eliminate the back-up problem.

The Drain King attaches to the end of your garden hose. With a body of rubber about the same

Wrightway's Little Pete and Big Pete augers allow you to feed the snake into a pipe merely by turning a handle.

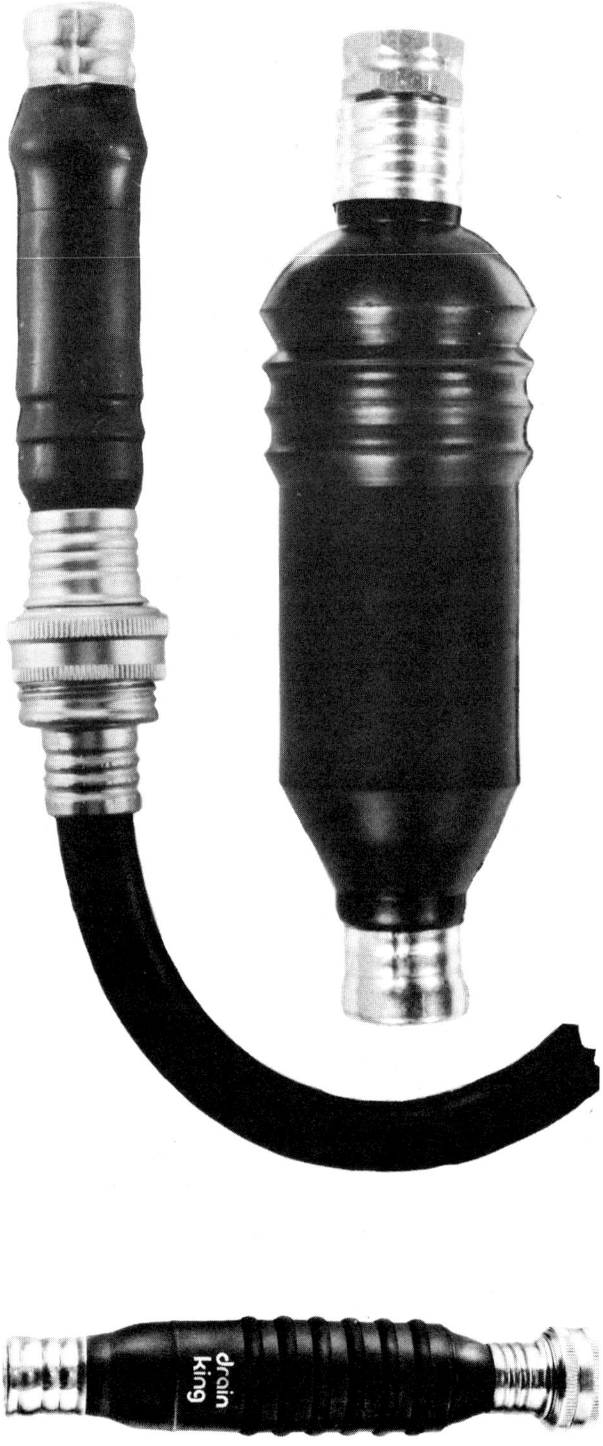

The Drain King unclogger comes in different sizes for different sized drains, but all work the same way and all are made to attach to the end of your garden hose.

diameter as your garden hose (it comes in different sizes for different sized drains), it can be inserted easily into any drain that does not have a screen or grid cover. Once you turn the water on, however, the Drain King expands and locks itself inside the pipe. Since the device completely fills the pipe, no water can come back out and all the pressure is directed against the clog. Moreover, there is a surge valve inside the Drain King that forces the water out in pulsating blasts that help break up the blockage.

Like any other pressure cure for clogged drains, the Drain King's pressure must be directed full force against the blockage. If there is an escape route possible, the pressure will follow the path of least resistance. Therefore, you must cover all the escape routes in the immediate area. In addition, if the blockage is beyond the vent pipe, the water you send in via the Drain King may just go out through the vent pipe and onto the roof. Then, when you turn off the water and release the pressure, all the water that is in the vent pipe will just drain back out toward you. If you feed the hose down a sink drain, then the problem is not a serious one; but if you remove a trap to get the hose into the pipe, then the water from the vent pipe will drain back out on the floor.

Sometimes, when you turn off the water, the Drain King stays expanded and refuses to come out of the drain. To remove it, however, all you have to do is disconnect the hose at the hydrant, and the Drain King will deflate. If you have a swimming pool, you will certainly want a Drain King for keeping the lines cleared.

TOILET TANK PARTS

Ball cock assemblies come in various price categories, ranging from around $2 to more than $30. There are inexpensive brass models and newer type plastic units. The best of the new plastic ball cock assemblies is the Fill Master, which works well and is easy to install. Although the Fill Master is too new to permit any conclusions to be drawn regarding its long-term durability, it should stand up well. Close behind is the Fluidmaster, which comes in various models. The less expensive models — 100 and 200 — work on the same principle, but Fluidmaster's top-of-the-line model 400 is sturdier.

Brand names are of no concern when you need to replace float balls, lift wires, and other small parts inside the tank. All makes are about the same. When you must replace a tank ball, however, there is a new product you should consider. It is called a flapper, and the two most popular makes are the Korky and the Water Master. Since a corroded seat is frequently the cause of tank ball trouble, Fluidmaster is now introducing a flapper that includes a stainless steel replacement seat. Thus, as you replace the tank ball, you also provide a corrosion-free seat for the flapper. The real beauty of this unit, though, is that the new seat is sealant mounted over the old;

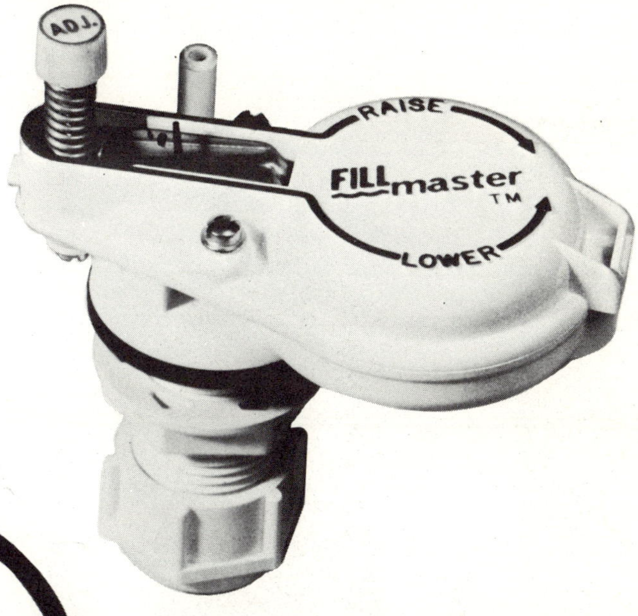

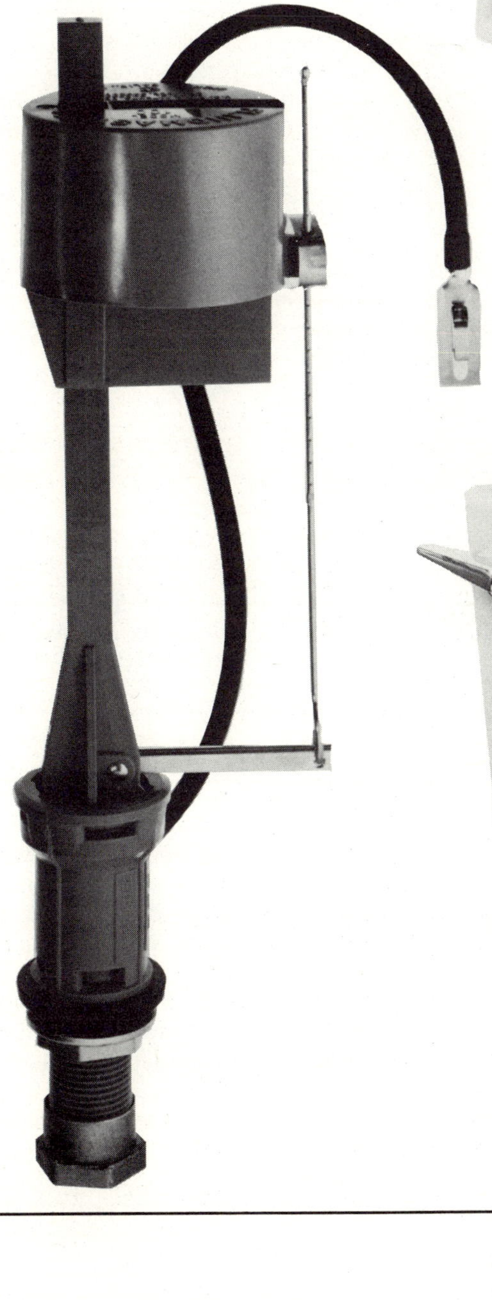

The best of the new plastic ball cock assemblies is the Fill Master (top, right). Close behind is the Fluidmaster: the less expensive model 100 (bottom, left) and the top-of-the-line model 400 (top, left). You can change the water level by sliding the Fluidmaster's cup (bottom, right).

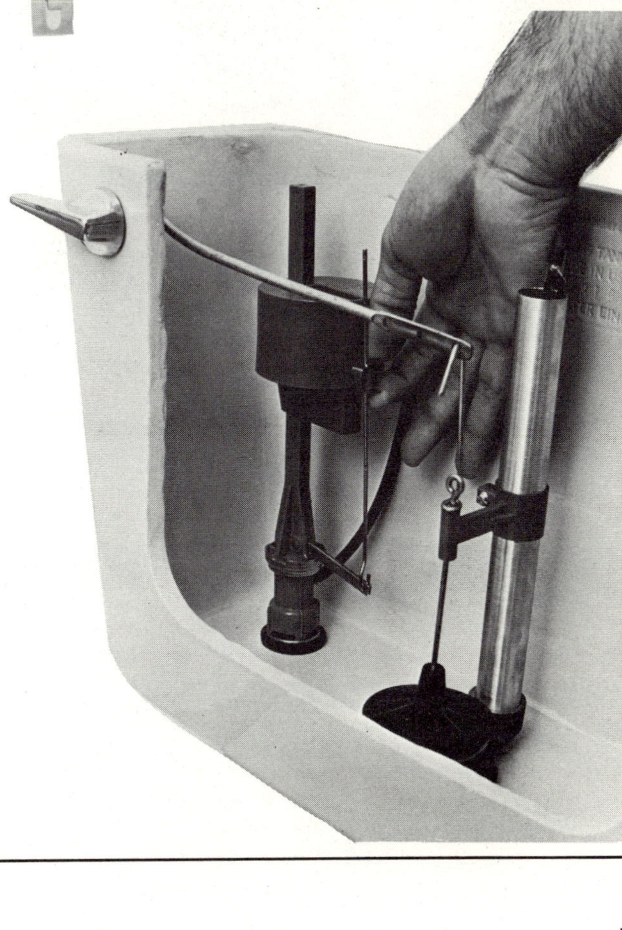

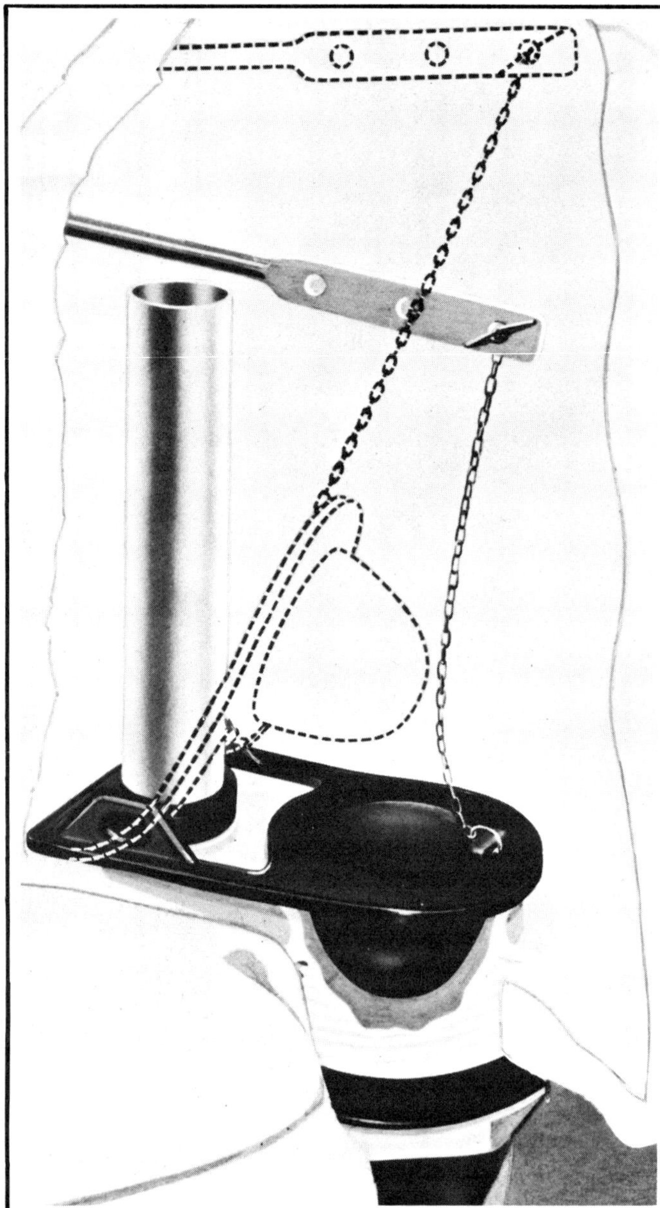

The Korky Flapper is one of the most popular replacement products for the traditional toilet tank ball.

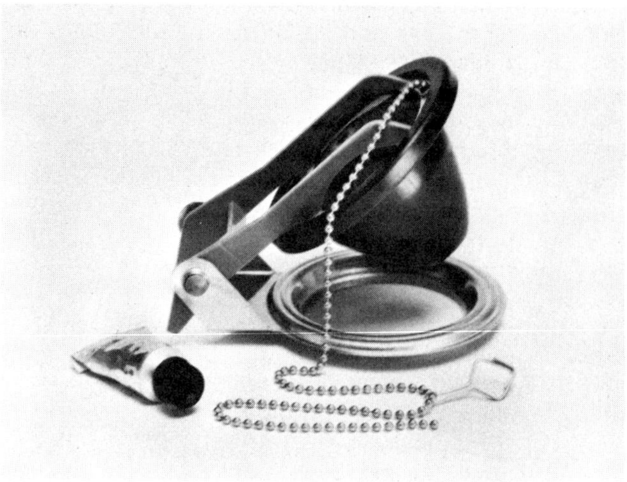

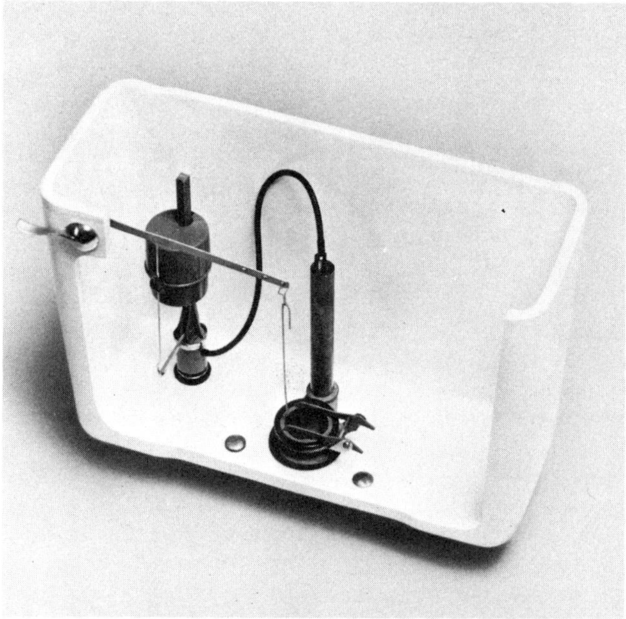

Fluidmaster's Flusher Fixer Kit includes a flapper with a stainless steel replacement seat (top). Best of all, though, you need no tools to install the flapper; it mounts with watertight sealant inside the tank (bottom).

therefore, you need no tools for installation. Fluidmaster calls its new unit the Flusher Fixer Kit.

PIPE FITTINGS AND WRAPS

When you get ready to add a cutoff, replace a faucet, or install a new trap under the sink, you will be surprised at how easy the plumbing supply manufacturers have made your work. Most hardware and plumbing supply retailers carry carded plumbing fittings, which means that each component is packaged separately on a card. When you turn the card over, you find detailed instructions for installing that particular fitting. Among the manufacturers who package their plumbing products for the do-it-yourselfer are EZ Plumb, Plumb-Rite, Plumb Shop, Harrington-Gross, and Chicago Specialty. Their products are of good quality, and their instructions are easy to follow. In addition, most of these companies make certain that retailers keep their stock of parts up to date, minimizing your chances of finding all but one part for your plumbing project. You may pay a bit more for such packaged parts than if you picked the same quality fittings from an open bin, but the full stock factor and the do-it-yourself installation instructions are worth the extra cost.

Most pipe wraps are designed either to provide freeze protection or to control pipe condensation. In the condensation control department, NoDrip, Wrap On, and Frost King pipe wraps all provide satisfactory results. Wrap On and Frost King also do a fine job insulating pipes against winter freezes. In addition, Wrap On's electric heat tape is highly recommended for any extremely cold weather situation.

Competing against the brand names mentioned, there are several private-label plumbing products that actually sell at a greater volume than the highly advertised brand-name products. Sears, Montgomery Ward, and J.C. Penney all carry their own private-label brands, and all stand behind their products. Avoid purchasing private-label plumbing products at retail stores that do not possess the reputation for quality merchandise which Sears, Montgomery Ward, and J.C. Penney enjoy.

SUMP PUMPS

Most people who are blessed or cursed with a basement know the panic that water rising in underground rooms can engender. In many cases, a sump pump can eliminate a great deal of the anguish associated with a flooded basement. There are two popular types of sump pumps: the upright and submersible. Both types are installed in a small well, called a sump well, which can be either a poured concrete affair or a section of glazed 24-inch sewer tiles set on end with a gravel base. When the water reaches a certain level in the sump well, it turns the pump on. If the pump is of adequate discharge capacity to handle all the water, the basement floor will stay dry.

The difference between the two pump types is that the motor on an upright is at the top of the unit and sticks up through the cover above the sump well, while the submersible unit does not protrude above the pit. Since the motor in a submersible pump is sealed, it can be covered with water. The upright pump, which is generally less expensive than the submersible pump, is activated by the rising water. The water carries a float upward, and when the float reaches a certain point, contact is made and the pump motor starts. As the water is pumped out, the float drops, eventually reaching the point that shuts off the pump. Since these points are adjustable, you can decide how much water you want in the well before the pump goes to work. The cutoff point is quite important because a certain level of water must be in the well at all times to act as a trap for sewer gases.

Submersible type pumps may be activated by a float or by air pressure. In the latter variety, the rising water compresses air trapped inside the unit; when the pressure reaches a certain point, the pump switch is engaged and the motor starts. When the air pressure drops to a certain level, the switch shuts the motor off.

Since the water level in the sump well is

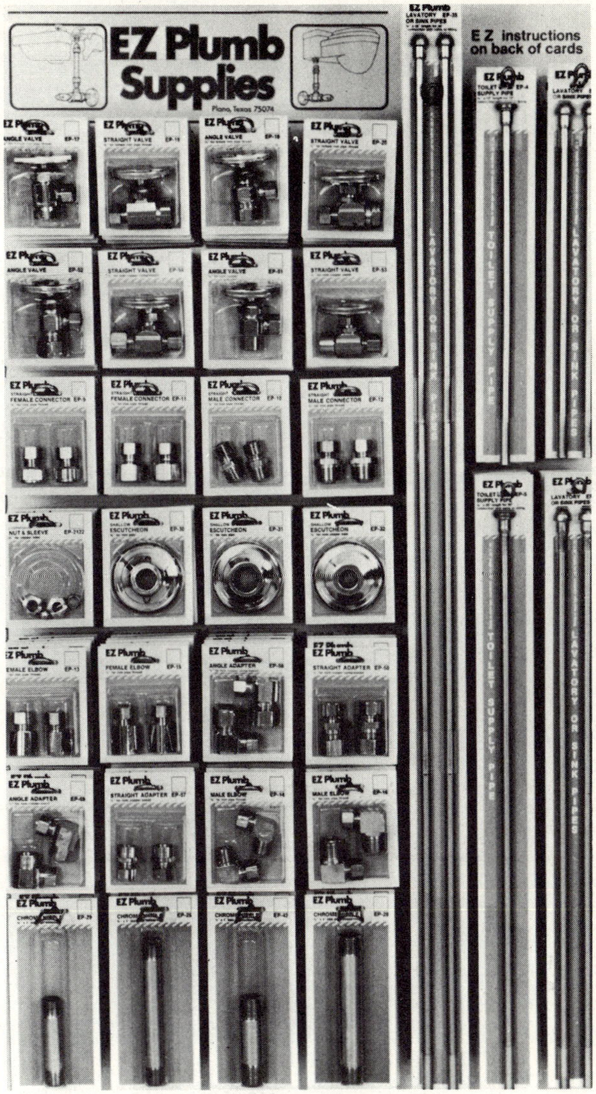

EZ Plumb is only one of several plumbing parts makers who offer each component packaged on a separate card.

Mortell's self-adhering NoDrip tape is designed to stop condensation (sweating) problems by insulating the pipes.

important as a trap for sewer gases, you should check the level regularly. As a matter of fact, it is a good idea to add water on a regular basis during intervals when there is no outside water going into the sump well. Just pour the water down the drain in the floor, and the water will make its way into the sump well on its own.

If you detect a sewer gas problem, it is probably due to the float being set incorrectly. Move the cutoff point up to make the pump shut off sooner and leave the standing water level a bit higher. A good rule of thumb is to leave about six inches of water in the sump well. Since some submersible pumps cannot be adjusted by homeowners (the starting and cutoff points are preset at the factory), you must take such a unit to the factory or a service center for setting. Therefore, consider future service requirements when purchasing a pump. If you buy a pump that is preset at the factory, make sure that there is an authorized service center convenient to you. You cannot afford to be without the pump for very long.

Regular cleaning of the sump well is a must. Residue can foul up the pump, clog up the lines, and may even destroy the possibility of electrical contact at the time the pump should go on. Before you remove the sump well cover, however, be sure to unplug the electrical cord and flip the circuit breaker switch or remove the fuse. The mention of electricity should bring to mind another possible problem with your sump pump. What happens if there is a power failure and your sump pump cannot do its job? You should have a reserve pump that does not depend upon electric power. Hydraulic pumps turn on automatically when the water reaches a certain level, and there is a very inexpensive unit that uses your home water supply pressure to siphon away water. The latter is connected to a pair of garden hoses, and it can carry away about four gallons of water for each gallon from the water supply. Since the unit has no moving parts, it cannot cause much trouble; but, of course, you must activate it. Other reserve pumps include the battery-powered models and the gasoline-driven pumps.

If you experience any backflow problem in the drain line from the pump, it means that the check valve unit is not functioning properly. The check valve has a hinge that opens when there is water being pumped through it. When the pump stops, the hinge closes, and back pressure keeps it closed tightly. The check valve stops the return of sewer gas. If the check valve unit on your pump is faulty, you can install a replacement without much difficulty.

When purchasing a pump, never settle for one that might be adequate. Buy a pump that might be a little too big. In addition, pay a little extra for the kind with safety devices and with overload cutouts that may save your pump motor. Be sure to follow the manufacturer's suggested maintenance schedule as described in the owner's manual. It is a good idea to check the pump regularly during a long dry spell to make sure that the machine is still operational. Although the installation of a sump pump and sump well may not be everybody's idea of a do-it-yourself project, putting in a replacement pump should be no problem. Give your sump pump tender loving care, however, and you should find that the old pump will last for many years.

WATER CONDITIONERS

The water supply in some communities contains minerals and chemicals in such abundance that it can cause all sorts of nuisances. Such water is referred to as "hard water," and it can cause deposits to form in your supply pipes so that the flow of water is reduced. Hard water can make it impossible to get your clothes truly clean, force you to use more detergent, leave rings around your tub (as well as around your collar), streak your glassware and dishes — it can even make your drinking water taste, smell, and look unpleasant.

It is impossible to recommend specific types of water softener units because the best type for you depends upon the nature of your water supply. People who sell water conditioners generally test water free. You can learn from the tests which minerals in particular are causing the hard water problem; then, you can determine what size and type of softener and/or filter you should install.

Most water softeners operate by substituting soft ions for the hard ions. Since the soft ions are sodium based, any member of the family on a salt-free diet should be provided with a separate unsoftened outlet. Many hardness conditions affect only the hot water, moreover, and your unit may operate only on the hot water lines. Outside hydrants and toilets, of course, need not be equipped with softened water, and therefore, it is very possible that only specific parts of your home's water supply system will be run through the water softener.

In some cases, the water must be conditioned with the addition of chemicals into the water. A device called a feeder injects small amounts of chlorine into the water. If you actually need to remove elements from the water, you must install a filter. Different types of filters are required for the removal of different kinds of minerals and for the removal of different quantities of minerals. If your water supply is full of deposits, you may need a softener, a feeder, and a filter.

Generally speaking, whatever apparatus you need to condition the water supply can be installed on a do-it-yourself basis. Your dealer can be very helpful by informing you as to what fittings you will need and how to position the unit so that it affects only the portion of the water supply that needs to be treated.

Most types of water conditioners are easy to service. For example, when all the sodium ions have been consumed in a softener unit, you can treat the media with a salt solution to regenerate it. Filters can

Redyhot is a compact unit that provides hot water for drinks and soup instantly at the push of a button. Most of the appliance is concealed beneath the kitchen sink.

either be cleaned or replaced. In areas where hard water is a problem, you can often rent a conditioner. Renting eliminates installation and servicing, but it also eliminates more of your family budget. Therefore, you owe it to yourself to at least consider water conditioning as a do-it-yourself project.

PLUMBING MISCELLANY

There is a new gadget, called a Redyhot, that provides instant hot water for those who cannot wait for the pot to get hot. Redyhot is a compact unit that ties into your water supply and your house electric current. It can be installed under the sink, with only the unit's gooseneck and push-button activator showing above the sink. Installation is simple enough for any do-it-yourselfer, and you can put in a Redyhot without any special tools. If you like convenience, you will enjoy a Redyhot.

People who wrap their outside faucets every year with outlandish old rags to protect the faucets against Jack Frost will rejoice at knowing that there are new plastic faucet covers. The dome-shaped covers are designed for use on hydrants that come out from the wall. Made from polystyrene (a plastic with excellent insulation qualities), the covers are easily installed and quickly removed because they are held in place by strong rubber bands from inside the cover. You merely stretch the rubber bands around the faucet to pull the dome down tight against the wall. The lip of the dome includes a gasket of foam rubber so that when it is pulled against the wall, the lip conforms to any uneven — like brick — surface.

If you plan to spend much time in your basement, you probably would like to add a toilet down there. Frequently, however, the sewer line is situated some distance above the level of the basement floor. If you are faced with this arrangement, you will be happy to know that there are actually toilets that flush upward. These units need no pipes or openings coming out of the basement floor, and the installation of an up-flusher is generally a do-it-yourself project.

The principle behind toilets being able to flush upward is a special flush valve. The valve first sends water through a jet that breaks up solids, and then it switches the water to another outlet that flushes the bowl and carries the now liquified matter up through a pipe and into the regular gravity drain. Normal city water pressure is adequate for an upflush toilet, even if the unit must flush upward a distance as high as the full basement height. A water pressure of 40 psi can flush waste from the unit 10 feet upward.

The toilet kit comes with all the fittings, pipes, and tubing you need to install it. All you have to worry about is getting the water supply to the location. The flush valve connects to your water supply line (ideally behind the wall), and the handle attaches to the valve on the outside of that wall. You anchor the bowl itself to the floor by whatever method best suits the type of floor you have. The best time to install the upflush toilet is when you are remodeling your basement. At that time, you can make all of the connections easily and then wall over all the pipes and the valve. Easy-to-understand instructions come with the kit.

Sometimes, you can prevent your basement from flooding with the addition of a special unit in your floor drain. Insert this unit (sold under the brand names Drain Guard and Flood Guard) into the drain opening, and secure it by turning a few screws to expand the device against the wall of the drain. If your drain is threaded, install one of the threaded models that you screw into place. When there is a backup in the pipe, the water lifts a float, and the unit seals the opening. When the water drains out, the unit just sits there. A Drain Guard or Flood Guard may not solve the really big basement flooding problems, but such a unit sure can stop plenty of little ones.

Chapter Seven

Hot Water Systems

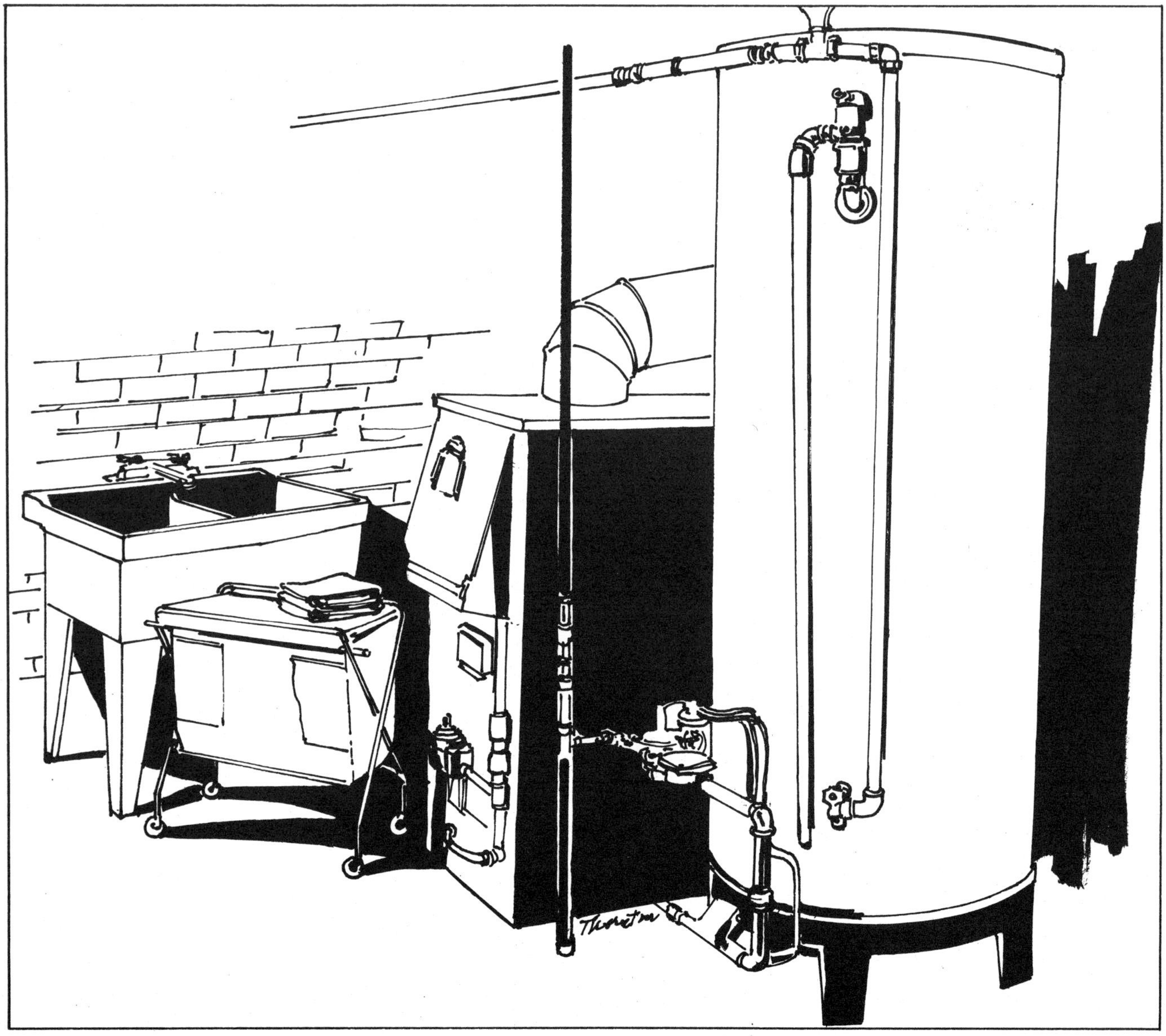

The key to having a trouble-free hot water system is to select a high quality hot water heater for your home. Of course, if you already have a water heater, this advice comes too late. Yet, by knowing how to select the proper heater, you may be able to understand some of the problems you are having with the one that is presently in your home. Besides, if your problems are really bad, you may be shopping for a new heater in the very near future. Keep these four factors in mind whenever you consider the purchase of a new hot water heater: (1) Size of the tank; (2) Warranty of the tank; (3) The tank lining; and (4) Recovery time.

Many times, people think that their old water heaters are broken because there never seems to be enough hot water. It may be, however, that the size of the tank is simply not adequate for the size of the family. "It used to do a great job," you say, but stop and think of how your hot water needs have grown. Dishwashers, laundry units, small children — all require plenty of hot water. The tank that was adequate a few years ago may now be overtaxed.

Planning, nevertheless, can expand your present heater's efficiency. Avoid a waiting line at bath time. Change some habits so that the family's baths are spread throughout the day, or some at night and some in the morning. Moreover, quick showers save on hot water. A shower consumes from four to eight gallons of hot water, while a tub bath can use from 10 to 12 gallons.

Arrange to use the automatic dishwasher and the clothes washer at different times of the day, and certainly do not run them at the same time people are bathing. Avoid running either machine with half loads, and rinse dishes with cold water before you place them in the dishwasher. Many times, we use hot or warm water when cold water could do the job. Modern fabrics are actually designed for cold water washing, and kitchen disposers certainly do a much better job with cold water. In addition, you can often use warm water instead of hot water for a great many household chores.

In most cases, you will be able to find 7½- to 15-year warranties on water heaters that are competitively priced. Since it generally pays to opt for the top of the line in water heaters, shoot for one that offers a long warranty. Remember, though, that the advertised warranty usually refers only to the tank, and that a meaningful warranty must be backed by a reliable manufacturer and dealer.

The most popular linings for water heater tanks are glass. Glass linings last much longer than other types, and they generally provide cleaner water. Copper-lined tanks are also good and long-lasting. Galvanized tanks, though way down on the popularity list, can be satisfactory in localities where the water possesses few of the harmful elements that could corrode a galvanized tank. Certainly the least expensive in purchase price, a galvanized tank can prove to be quite expensive if you must replace it after only a couple of years.

In the hot water heater jargon, recovery time usually means the number of gallons per hour that the heater can raise 100 degrees. Be careful, however; some manufacturers show the recovery rate of their heaters based on a 60-degree rise. A high recovery rate is very important for families whose schedules force them to consume a great deal of hot water at the same hours of the day.

The only other aspect of purchasing a new water heater that you should consider is the method of heating. If you are replacing a unit, it is almost always preferable to stay with the same type of fuel: gas or electric. In a new installation, the cost and availability of gas versus electricity should be your primary considerations. In most cases, a natural gas hot water heater is less expensive to operate than an electric hot water heater.

If you plan to make the installation a do-it-yourself project, then be sure to check with the local

Matching Water Heater Size to Family Needs

		Gas				Electric			
		Bathrooms				Bathrooms			
	No. of People*	1	1½	2	3	1	1½	2	3
No dishwasher or clothes washer	2	30 gal.	30	40	40	30	42	42	42
	3	30 gal.	40	40	40	42	42	42	52
	4	40 gal.	40	50	50	42	52	52	66
	5	40	50	50	50	52	66	66	82
	6	50	50	50	50**	66	66	66	82
With either dishwasher or clothes washer	2	40	40	50	50	42	42	52	52
	3	40	50	50	50	42	42	52	52
	4	50	50	50	50**	52	52	52	66
	5	50	50	50**	50**	52	66	66	82
	6	50	50	50**	50**	66	66	66	82
With both dishwasher clothes washer	2	40	50	50	50	42	52	52	66
	3	50	50	50	50	52	52	66	66
	4	50	50	50**	50**	52	52	66	82
	5	50	50	50**	50**	66	66	82	82**
	6	50	50	50**	50**	66	66	82	82**

*Count each child under 10 as two people
**You might do well to consider two units

building code to see whether there are restrictions as to the type of water heater allowed, and — more importantly — whether you can legally perform the installation yourself under the provisions of the code. Some communities require that a licensed plumber install your new hot water heater.

HOW THE HEATER WORKS

The water heater consists of an insulated tank with a source of controlled heat (electric elements or a gas burner), surrounded by a cabinet. If the supply line enters the top of the tank, an internal tube — called a dip tube — carries the cold water to the bottom of the tank. In some heaters, though, the supply line enters the bottom of the tank directly. Since cold water is heavier than hot water, it tends to settle near the bottom. The heated water, in contrast, rises toward the top of the tank. Since the hot water line always exits from the top of the tank, you always get the hottest water available. This construction method also explains why you never run out of hot water gradually. It happens suddenly as the tank empties and the cold water reaches the top of the tank.

Within the tank there is a magnesium rod — called an anode — which is placed there to prevent the erosion of tank parts that may not have been completely treated at the factory. In locations where the water supply is very acid, the anode itself can be eaten away within a few years time. Therefore, the anode is usually accessible through a plug in the top of the tank for easy replacement.

A bimetal thermostat controls the heat on an electric water heater by turning the elements on and off as required to maintain a preset temperature level. The thermostat is adjustable, with a dial scale on the front, although in many instances you have to turn off the power and remove the access panel to reach the dial.

In some cases, the dial indicates temperatures in actual degrees, while other dials just indicate warm, normal, or hot. Normal (or medium) should be the equivalent of about 140 to 150 degrees Fahrenheit. Always turn the power off before adjusting a thermostat or servicing the heater. Set the thermostat high enough to maintain proper temperatures, and allow for a temperature drop of about one degree per running foot of pipe between the heater and the fixture outlet. Your automatic dishwasher requires hot water of at least 140 degrees in order to clean your dishes properly, though a temperature setting of around 150 degrees is most desirable.

Electric water heaters generally have elements at top and bottom, bolted to flanges outside of the tank.

Gas water heaters must be vented to the outside. Often, the flue goes through the center of the tank to the vent.

When electricity is the power source, the tank possesses heating elements to provide the heat. In most newer heaters, the elements are located within the tank itself. They are bolted onto a flange that is on the outside of the tank, and a gasket seals the elements against water leakage. Some electric heaters have wrap-around elements; these are metal sheaths which are strapped tightly around the outside of the tank. Wrap-around elements heat the water throughout the tank.

Most electric heaters have elements and thermostats at top and at bottom. The top element begins to heat first, providing somewhat faster recovery because it is closest to the top of the tank. Since the upper and lower elements are interlocked, the lower element cannot go on until the upper thermostat has turned off the upper element. Then, the lower element turns on, heats the remainder of the tank, and shuts off when the water reaches the correct temperature.

On many electric heaters, a safety interlock is provided. Usually located near the top of the tank and often mounted above the top thermostat, the safety interlock is a manual-reset device to turn off the power to both elements if the water heater should overheat. To reset the safety interlock, you must turn off all power to the heater, remove the front access panel, and push a button. If the device should kick out a second time, however, be sure to locate and verify the reason. You can use a thermometer at a nearby faucet to determine whether the water is actually overheating.

Gas water heaters also have a thermostat to control temperature levels. The thermostat may be a switch, usually with a tube-type sensing device, that opens and closes a set of electrical contacts; or it may be coupled directly to the valve itself, using the motion of the expansion and contraction of a temperature-sensitive bellows to open and close the gas valve directly.

With the gas unit a pilot is lighted at all times, and the flame from the pilot is directed against a device called a thermocouple. As long as the pilot is on, the thermocouple sends a message to the gas valve that everying is operating as it should. Were the flame to go out, however, the thermocouple would cool and send a message to the gas valve that the pilot was out. As a result, the gas valve would shut off so that no gas would go to the burner when there was no flame to ignite it.

What causes a pilot to fail? It could be the wind blowing down the vent, or it could be caused by lint collecting near the pilot orifice. If your pilot should fail, look closely around the pilot area before relighting the flame. If lint has built up, use an old toothbrush to clean it away.

The most common cause for pilot failure, however, is a faulty thermocouple. Thermocouples are easy to replace, but you must be sure to get the proper kind that has a tube of the correct length.

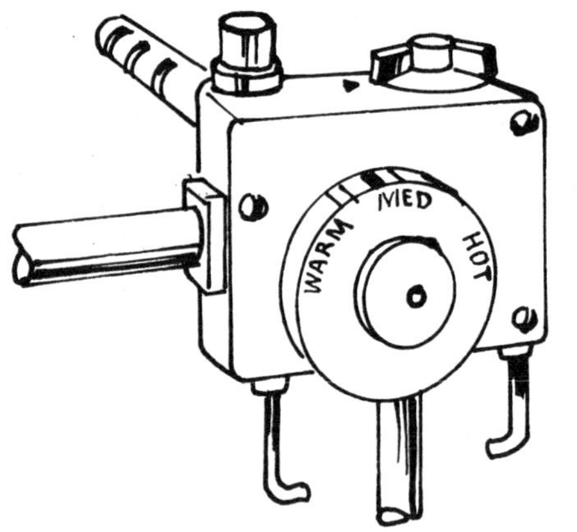

Some water heater thermostat dials indicate degrees, while others just indicate warm, medium, and hot.

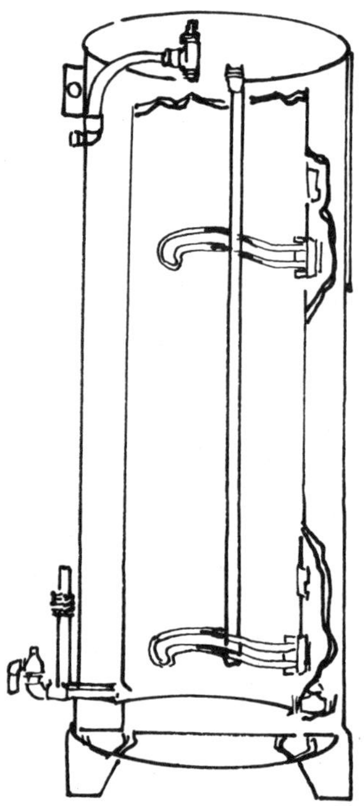

The top element in an electric water heater goes on first; the bottom one starts only after the top one shuts off.

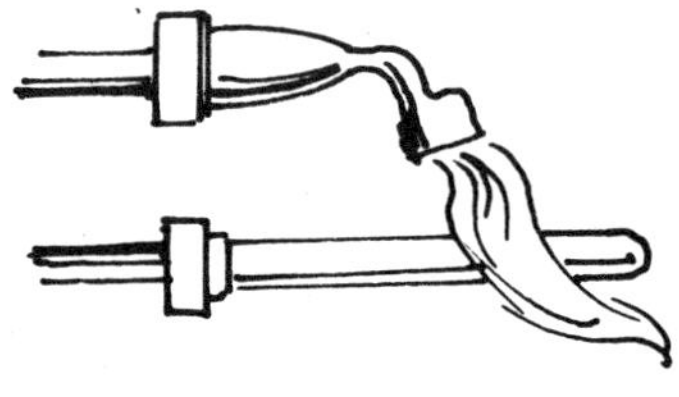

The flame from the pilot light on a gas heater bathes the top half-inch of the thermocouple safety device.

Usually sold carded and sealed with plastic, thermocouples come with installation instructions on the back of the card. Position the sensing device so that the pilot flame is directed on about the top half inch of the thermocouple's tip.

Gas water heaters must be vented to the outside of the home. Generally, the flue from the burner passes through the center of the water heater tank to the point where the vent is attached. Baffles within this flue help to divert much of the escaping heat to the interior of the tank where it can be used to heat the water.

In addition to the water heater safety controls already mentioned — the high-limit protector found on most electrical water heaters and the thermocouple on gas water heaters — there is one device that all water heaters should have: a combination temperature-pressure relief valve. The relief valve itself is attached to the top of the water heater tank because that is where the hottest water collects, while the valve's sensing element must be located in the uppermost portion of the tank. If either the pressure or the temperature within the tank becomes excessive, the valve opens and allows cool water to flow in and maintain a safe temperature level.

Never underestimate the importance of the water heater's relief valve. With an inoperative valve or with no valve at all, the water heater tank can build up extremely high temperatures and pressure; it can even explode under some conditions, and houses have been leveled by just this one factor. If your water heater does not have a safety valve, turn off the heater and do not turn it on again until you have a relief valve installed. Your heater should also have a drain line to carry excess water away from the tank; water expands greatly when heated, and some overflow — one or two gallons a day — is normal. Make sure, however, that your drain line does not go outside since a hard freeze could plug it.

MAINTENANCE AND REPAIRS

The water heater is one of the most important of the major appliances in your home. It seldom gives any trouble, but when it does, everyone in the household knows about it. Many of the problems that you encounter with a hot water heater are not generally the kind of repairs for the home handyman, but you will have to decide for yourself whether you can tackle such repairs or whether you should call in professional help.

Never attempt to make repairs on the burner of a gas water heater. Not only is special equipment required, but there is also a safety factor involved. Only an authorized technician should repair your gas burner. Consult your local utility company, and seek their advice when you trace a water heater problem to the burner.

If you want to make internal repairs on an electric hot water heater — such as element replace-

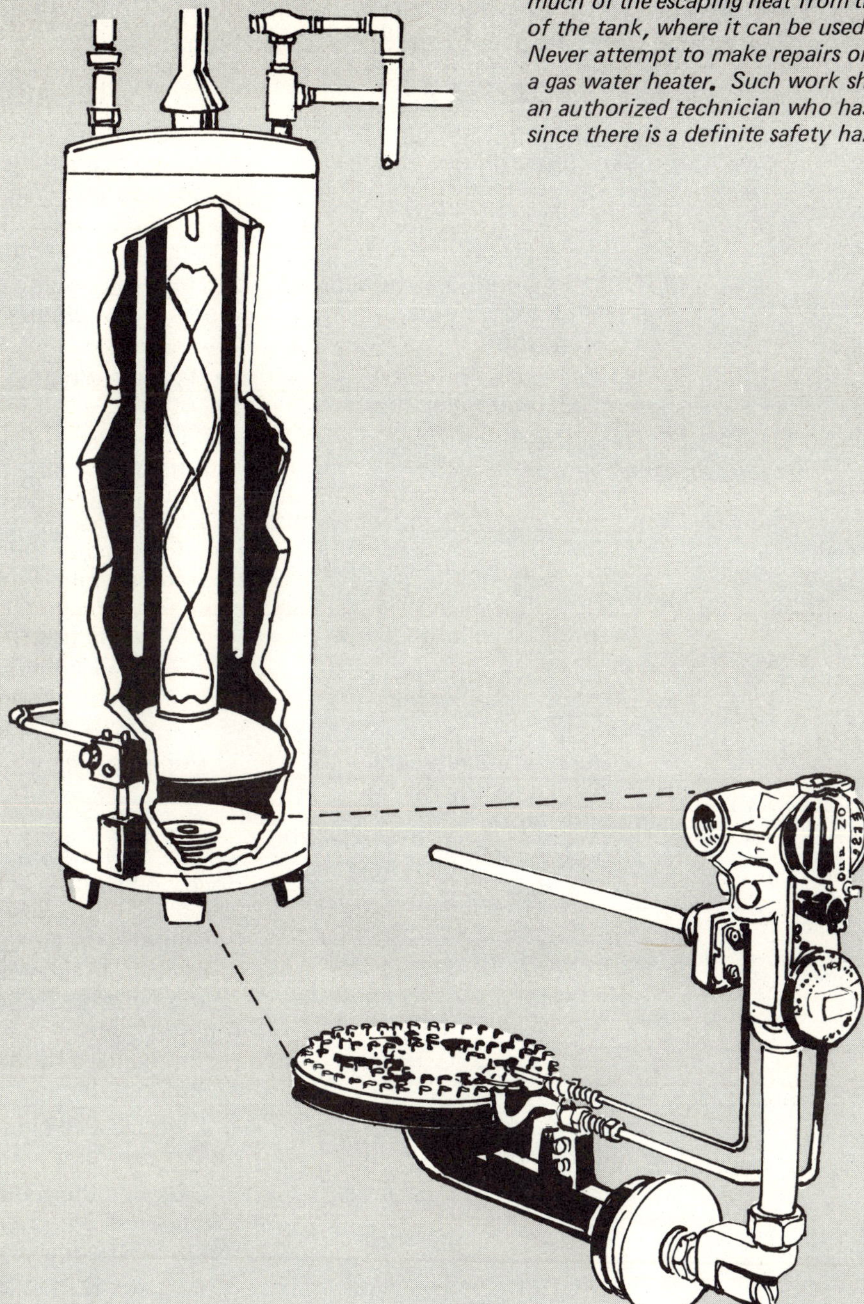

Baffles within the flue of a gas water heater help divert much of the escaping heat from the burner to the interior of the tank, where it can be used to heat the water. Never attempt to make repairs on the burner section of a gas water heater. Such work should be done only by an authorized technician who has the proper tools, since there is a definite safety hazard involved.

ment — be certain that all power to the water heater is turned off before you begin the job. If the water level should fall below the heating elements, the elements would burn out very quickly. Therefore, it is a good idea to turn off the water heater any time you cut the water supply in your home.

What can you do if you hear strange sounds emanating from the tank? The two most common causes for such noises are steam and sediment in the tank. You can frequently correct steam problems by merely turning the thermostat down (the normal setting is around 140 to 150 degrees), but you may have a faulty thermostat that no longer shuts off the elements or burner when the water reaches the desired temperature. To check the thermostat, turn the setting all the way down. If the heat source does

Water Heater Troubleshooting Chart

Turn off all electrical power or the gas supply before servicing, adjusting, or repairing a water heater.

PROBLEMS	CAUSES	REPAIRS
No hot water	1. Blown fuse or tripped circuit breaker (electric heater).	1. Replace fuse or reset circuit breaker. If same thing happens, a short circuit exists. Call service technician. Do not activate circuit again.
	2. Calcium buildup on heating element (electric heater, immersion type).	2. Have technician remove element and clean.
	3. Upper heating element burned out (electric heater).	3. Replace element.
	4. Defective thermocouple (gas heater).	4. Clean pilot orifice and relight pilot. If pilot refuses to stay lit, replace thermocouple.
	5. Defective thermostat.	5. Replace thermostat.
	6. High temperature limit reset kicked out.	6. Reset limit protector.
Insufficient hot water	1. Thermostat setting too low.	1. Normal setting is 140 to 150 degrees Fahrenheit.
	2. Tank too small for family's needs.	2. Install a larger unit.
	3. Lower heating element burned out (electric heater, immersion type).	3. Replace element.
	4. Defective dip tube.	4. The dip tube may have developed a hole, which is allowing cold water to dilute hot water near the top of the tank. The dip tube may have to be removed and replaced. If so, replace it with one recommended by the manufacturer.
Slow recovery	1. Heating element has calcium buildup.	1. Remove and clean the heating element.
Steam in hot water	1. Thermostat contacts stuck (electric heater).	1. Examine thermostat for shorted or burned terminals. Replace if necessary.
	2. Thermostat does not shut burner off (gas heater).	2. If burner fails to go off when thermostat is set lower, replace thermostat.
	3. Thermostat setting too high.	3. Normal setting is 140 to 150 degrees Fahrenheit.
Water beneath tank	1. Leaky element.	1. Replace element gasket or element.
	2. Leaky plumbing connections.	2. Make neccesary repairs.
	3. Condensation	3. Not much you can do except place pans underneath to catch drips. Not a serious problem.
	4. Hole in tank.	4. Replace tank.

not go out, then replace the thermostat. Thermostat failure in an electric heater can result from a short; check the terminals and replace the thermostat if you detect signs that it has shorted out.

There is no reason for sediment buildup ever to cause noise — if you check and drain the tank regularly. All you need to do is open the tap of the drain valve (it looks like an outdoor hydrant) at the bottom of the tank. Drain off a little water into a container. If it runs clear, stop. If it is not clear, keep draining small amounts until the water does run clear. After a few months, you will be able to determine how often you must follow this procedure. Not only does draining the tank eliminate noise problems, but it also allows your heater to operate more efficiently.

If your heater is an older model and you have never drained it, your drain valve may have become frozen. If so, do not force it. It is far better to hold off on draining the tank than to break the valve and flood your home. If your drain valve is frozen, you may have to replace the drain cock — a job which requires shutting off both the heater and the water supply, and then draining the tank before you can remove the faulty valve.

Sediment in the tank may be the reason why you see discolored water coming out from the hot water taps. There are some heaters that have a magnesium core inside the tank so that corrosive elements in the water attack the magnesium instead of other parts of the tank. When the rod has been attacked for some time, though, it can become so corroded that it releases particles that color the hot water you receive at the taps. If you have such a problem, you can usually eliminate it by replacing the magnesium core.

Sediment buildup can also slow down your water heater's recovery rate. If you notice a marked drop in your heater's recovery speed, be sure to drain the tank. In an electric water heater, the upper heating element can burn out and slow the recovery rate considerably. Replace the upper element if replacement is so indicated.

When there is no hot water at all, obviously the heat source is out. The first thing to check in a gas unit is whether the pilot is still on. If it is not, relight it. If the pilot is still lighted, make sure that the valve is turned to "On." Then turn the temperature control knob to its highest position. Turn on a hot water tap full force, and if the burner does not come on very shortly, your thermostat either needs cleaning or replacement. Usually the control knob slides straight off, and you can take off the cover from the thermostat unit by removing a few screws. If you see dirt or corrosion in the thermostat unit, you can usually clean it with a blast of air from a plastic squeeze bottle. If no dirt is present, then you probably have a faulty thermostat that must be replaced.

The first thing to check when an electric heater fails to produce hot water is whether a fuse may be blown or a circuit breaker switch tripped. If so, correct the electrical circuit. If the heater unit has a short circuit, then the fuse will blow again or the circuit breaker will trip again. You must track down the short. Look first at the wiring because the short may only be a loose wire.

If your electric heater has but one element, the problem could be that the element has burned out. With two-element units, however, it is unlikely that both elements would go simultaneously. You can put in a new element yourself, but be sure to cut off both the power supply and the water supply before you start draining the tank. In some cases, after you remove the element you discover that it is covered with calcium deposits. If that is what you find, you can clean the element with vinegar and a small brush, and then reinstall it. This procedure may not work, but it is worth trying before you buy a new element.

If your utility bills are too high, and you suspect that the hot water heater is the culprit, check to see whether you have a leak in the plumbing system. Many times, a leak can go unnoticed for a long time when it is under the house or inside the walls. Of course, a hot water faucet that leaks can be wasting a great deal of your money every month.

If the tank springs a leak, you can repair it; but you should start shopping around for a new unit. A leak in one spot generally indicates that the tank is rotting, and that a new leak will occur after you patch the present one. Patching a water heater tank is usually accomplished with a boiler plug that is inserted into the hole and turned down tight. Sweat soldering or tightening joints usually can put an end to leaking connections.

Many people complain about the fact that some of their hot water faucets run cold water for what seems like an eternity before any hot water comes out. This situation is not the fault of the hot water heater. It results from the fact that the heater is so far away from some fixtures that it takes time for all the cold water in the pipes to leave so that the hot water from the tank can come out. If you are really disturbed by having to wait for hot water, then you can have a circulating hot water system installed; such as system keeps hot water at the tap at all times. The best time to have a circulating hot water system installed, however, is when your house is being built. Another aid in getting hotter water to the tap faster is to wrap the pipes with special insulating tape; the tape contains aluminum foil to stop the pipes from losing heat. The best time to wrap the pipes is, of course, before they are enclosed in the walls.

Fortunately — because we depend on it so — the hot water heater is among the longer lasting appliances. These hints can help it last even longer. All of this information is general in nature, however, and the best guide for troubleshooting your water heater is the owner's manual you received at the time you purchased the unit.

Chapter Eight

Private Sewage Systems

People who live in an area where there is no municipal sewage-disposal system must have private sewage systems. The septic tank is now the standard equipment for a private waste-disposal facility, having almost totally replaced the old-fashioned cesspool. Both septic tanks and cesspools, however, are generally inferior to most municipal sewer systems.

TYPES OF SYSTEMS

There are two types of cesspool systems: the basic version is the leaching cesspool, while the second type employs a watertight cesspool tank in addition to the basic leaching unit. The leaching unit, sort of a dry well, has porous rock sides and bottom to permit the liquid matter that is drained into it to seep out into the surrounding earth. The earth is a natural filtration system; as waste flows or seeps through, the earth removes all of the waste matter's impurities. The more gravel and sand in the subsoil, moreover, the better the filtration is. After considerable usage, though, the solids in the sewage waste clog up the porous walls, preventing the liquid from seeping through. When that happens, the cesspool overflows and creates an unsanitary situation. Many systems, therefore, include a second cesspool to handle the overflow.

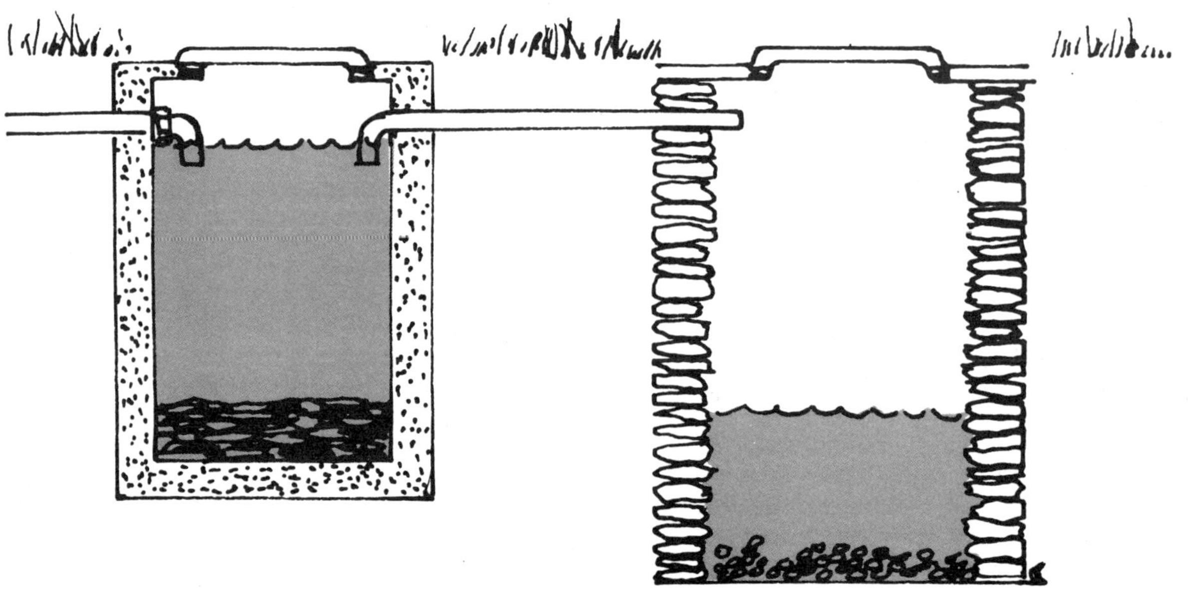

The better type of cesspool employs two tanks: the first is solid and watertight to retain nonliquid waste, while the second is a leaching tank that permits the liquid waste to seep out into the earth for purification.

The other type of cesspool system employs two tanks. The first tank, solid and watertight, retains the nonliquid waste matter, and allows the liquid to flow into the second tank. The second tank, a leaching tank, lets the water seep out for filtration. Since the first tank traps all the solids, there is little chance that the spaces between the rocks in the leaching tank wall can get clogged. Of course, the watertight tank must be cleaned out periodically by a professional firm that disposes of the solid waste.

Many communities have outlawed both of these cesspool systems. There are several good reasons for banning them, the principal one being that if there is any underground water nearby, the cesspool will slowly but surely pollute it. Naturally, such pollution makes the water dangerous to consume. In addition to the potential health hazard, cesspools can generate obnoxious odors as the waste matter works its way to the surface.

The septic tank, which is much better than the cesspool, is the only private sewage-disposal system still being built in this country. The septic tank provides a place where bacteria can go to work on the solids and convert them to liquid. The liquid is then dispersed into absorption fields, where it is purified as it filters through the soil. The bacterial action is a natural phenomenon, but the addition of certain chemicals (i.e., Bac-Tivator, Enzivator, FX4,

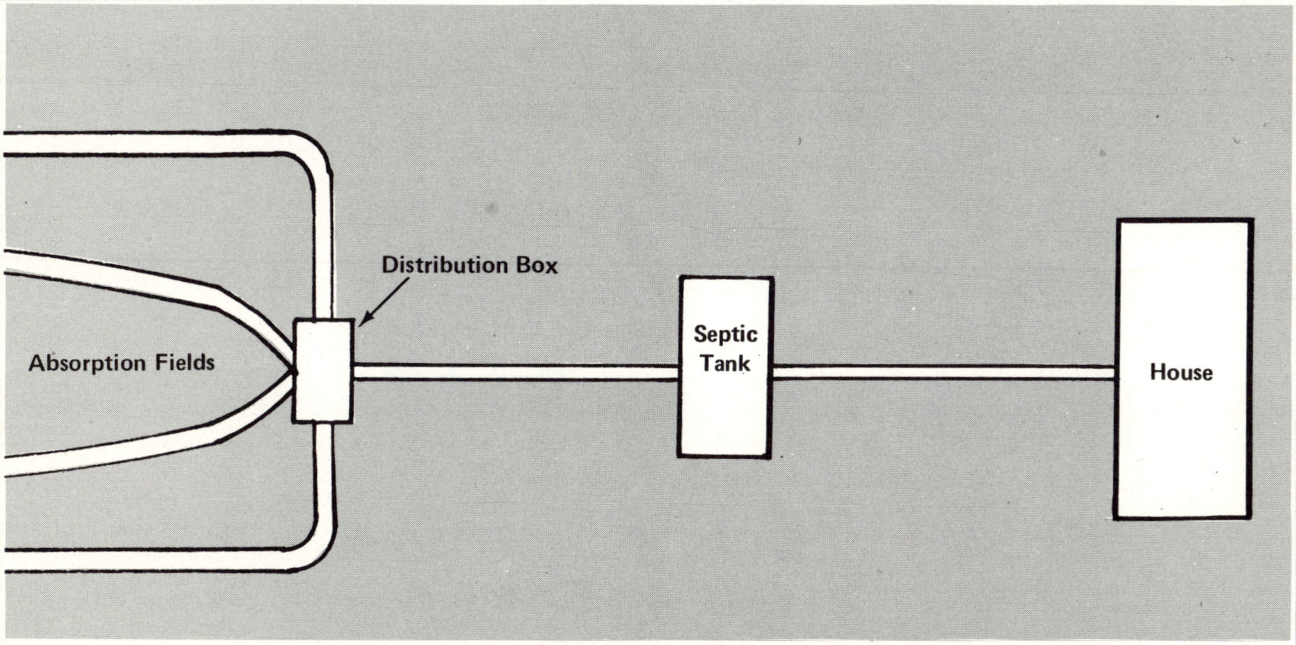

The septic tank system is better than either type of cesspool arrangement. Waste goes from the house to the tank, where solids are retained. Liquid waste then flows out to a distribution box and to the absorption fields.

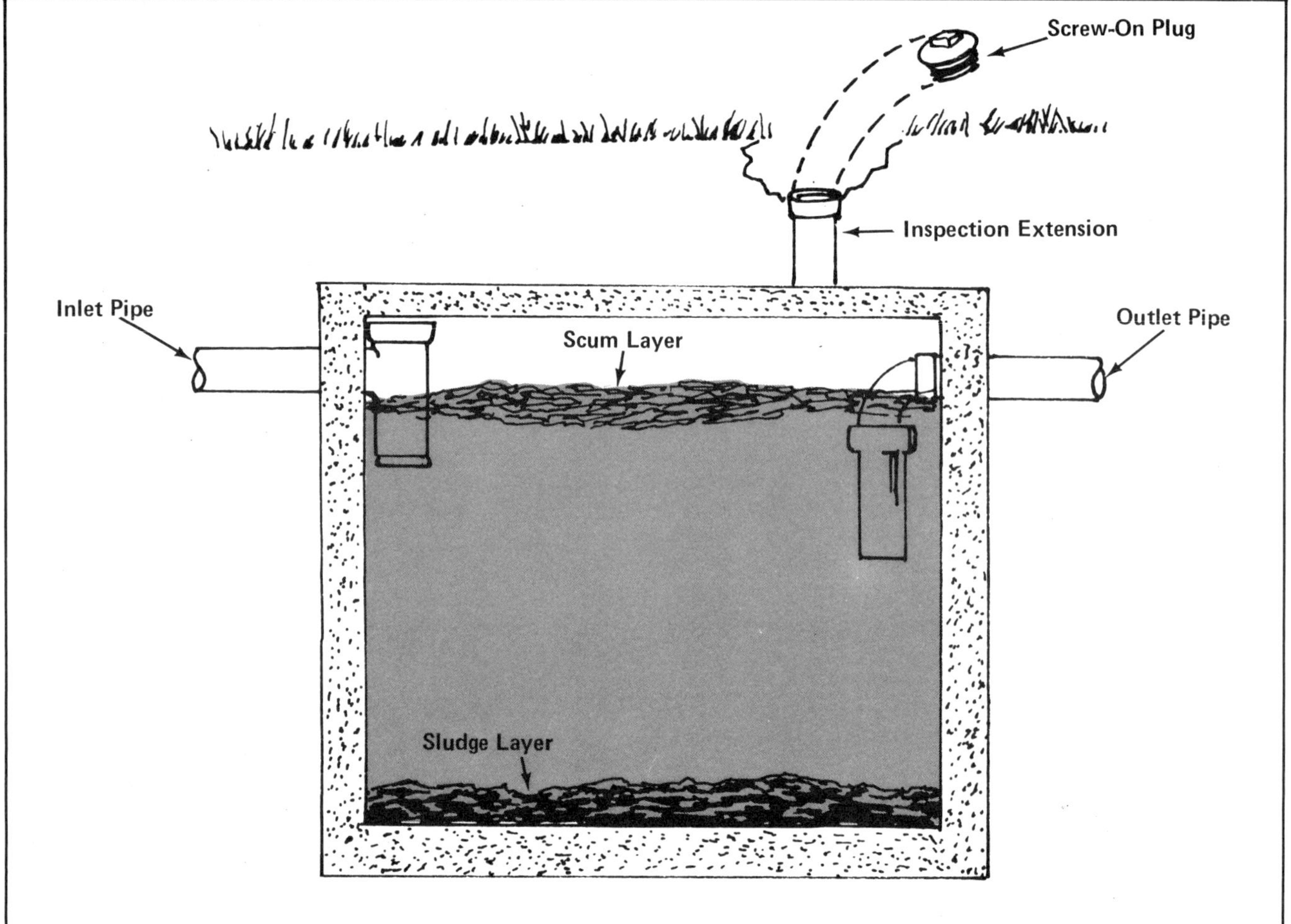

Although you can call in a professional firm to test your septic tank, there is no reason not to do the job yourself. Most tanks have an inspection extension, usually a pipe with a screw-on plug, that protrudes from the top.

Roebic Formula K37, Septic-Kleen, Sanzyme, Septic-Aide, and Septicare) into the system can speed up the liquifying process. What chemicals you use depends on the soil conditions and the type of sewage. You can also quicken bacterial action by introducing yeast into the septic system. Put about one half pound of brewer's yeast powder in a bucket of hot water, empty the contents of the bucket into a toilet bowl, and flush it into the septic tank system.

The sewage that is drained into the tank, of course, contains millions of bacteria which thrive and multiply in the absence of air and light. The bacteria attack all the solids, generating gas in the process. The gas rises to the top and carries small particles of solids with it to form a layer — called scum — at the top. Since both the inlet pipe and the outlet pipe on the tank are positioned below the scum level, the liquid can come in and go out without disturbing the layer of scum on top. From the tank, the liquid goes either to a drain field or into seepage pits, while the solids that escape the effects of the bacteria (or are destroyed very slowly) remain at the bottom of the tank. Every so often, this bottom layer — called sludge — as well as the scum layer must be cleaned out by a professional septic tank cleaning firm.

TESTING AND CLEANING THE TANK

A septic tank should be checked about once a year, but it may require cleaning only after several years (some tanks go as long as ten years between cleanings). It is most important that the layers of solids, both top and bottom, never get so close to the outlet that the solids can be carried out of the tank and into the drainage fields. The same firm that cleans out and removes the solid waste matter can check the tank on an annual basis, but the homeowner can make the inspection himself.

Most tanks have an inspection extension that comes from the top of the tank. Usually a pipe with a screw-on plug, this extension and its plug are buried only a few inches below the ground level. When you uncover and uncap the opening, however, you must allow the gas in the tank to disperse before you try to make the inspection. Do not breathe the gas; it can asphyxiate you. In addition, make sure

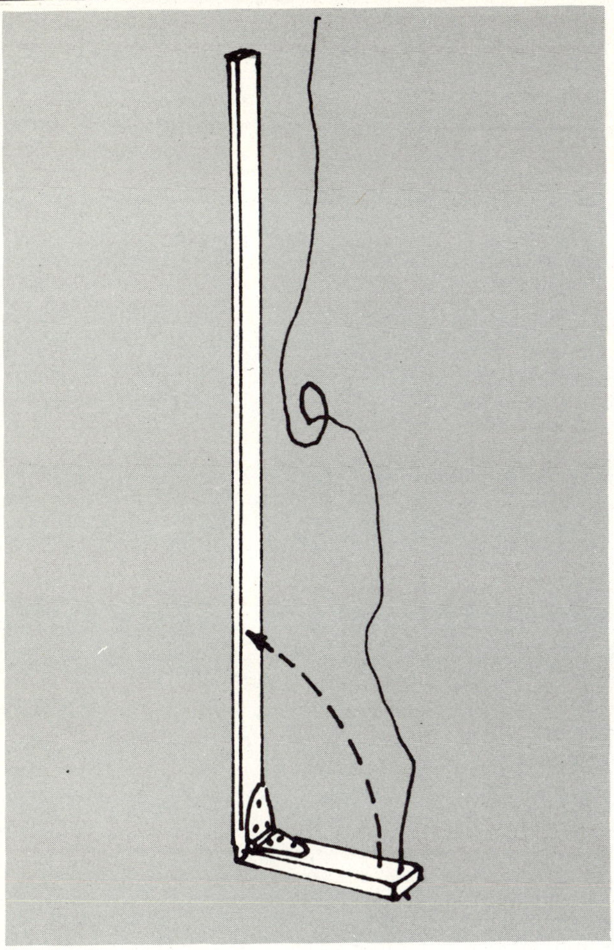

To make a scum test stick, attach a short lx2 to a long lx2 with a hinge, and then tie a string to the flap.

that there is no flame or spark around; the gas can explode.

You have to make a stick to test the scum level; fortunately, the scum test stick is easy to make. Attach a foot long flap piece of 1x2 inch board to a long 1x2 with a hinge. The hinge allows the flap to fold upward. The exact length of the long piece depends upon how deep your tank is buried. Position the flap right at the end of the long stick, and tie a stout string to it so that you can fold the flap and hold it next to the stick. Fold the flap up and insert the stick into the inspection extension, forcing it through the layer of scum. After it passes through the layer and beyond the top of the folded-up flap, release the string. The flap will fold down. Move the flap until it is against the bottom of the outlet pipe, and make a mark on the stick at the top of the extension pipe. Now, move the flap away from the outlet pipe, bring it up against the bottom of the scum layer, and mark the stick again. The distance between the two marks tells you how close the scum layer is to the bottom of the outlet pipe. If this distance is three inches or less, your septic tank needs cleaning.

The sludge test stick is nothing more than a long piece of 1x2-inch board with scrap cloth wrapped around the bottom portion. The length of the wrapping should measure the same as the distance from the bottom of the outlet pipe opening to the floor of the tank. Place the test stick in the inspection extension pipe, through the hole in the scum, and right on down to the bottom of the tank. As the stick goes into the sludge, gently turn it so that the end goes down to the bottom of the tank. The turning also causes the sludge to mark its level on the cloth. When you reach bottom, remove the stick slowly and measure the distance between the top of the sludge layer and the top of the cloth wrap (which is also the bottom of the outlet pipe). Then, consult the sludge allowable table to determine whether the sludge level is high enough to warrant cleaning. The cleaning firm will leave a small layer of the sludge in the tank to assure that the bacterial action will continue uninterrupted.

DISPOSAL FIELDS AND SEEPAGE PITS

The septic tank itself cannot purify the sewage. All it does is liquify some of the solid waste matter. The still-contaminated liquid leaves the tank, and it goes into either a disposal field or seepage pit

Sludge Allowable Table

Tank capacity in gallons	Depth of liquid in feet: 3	4	5
	Sludge test distance from top of sludge to bottom of outlet pipe in inches		
500	11	16	21
600	8	13	18
750	6	10	13
900	4	7	10
1,000	4	6	8

(the disposal field is usually the better of the two). The liquid goes through the outlet pipe to a distribution box, which is nothing more than a closed vat with several openings to allow the liquid to follow several different routes. From the distribution box, the liquid flows into the disposal or absorption field via pipes that are either perforated or loosely jointed. Released either through the perforations or the loose joints, the liquid waste seeps down into a bed of crushed stone or gravel, and a natural purification process begins.

Where there is not enough room for an adequate drainage field, seepage pits are the answer. Seepage pits, which are really nothing more than large dry wells, can also be utilized where the terrain is too steep for easy disposal field construction. The walls of the pits — concrete blocks, bricks, or stone — are not generally held together by mortar, and the liquid from the septic tank flows into the pits and seeps through the openings between the blocks, bricks, or stones. The bottom of the tank is filled with crushed rock, and sometimes, if the walls weaken, the entire pit fills with loose stones. The seepage pit system is similar to the watertight cesspool, except for the fact that the seepage pits receive waste that has first undergone the bacterial action of the septic tank.

CONSTRUCTION AND MAINTENANCE

Generally speaking, the local code, the state health department, or some other regulatory body specifies where the tank should go, how far it should be from your house, how far from your neighbor's house, how far from your water well (if any), and of what materials the tank and pipe should be constructed. In many areas, two separate septic systems are required. One is for toilet sewage, and the other is for all the other waste. There are excellent reasons for having separate systems. For example, the bleaches and detergents used for washing are strong enough to break down or slow down the bacterial action necessary for effective operation of the septic tank on toilet sewage. A washing machine itself puts far too much water into a septic tank. The sudden flow of water comes on so strong that it churns up the sludge, and some of the solids that should be dissolved by the bacteria are carried out through the outlet pipe and into the disposal field. The result — clogging — can cause unsanitary conditions.

There are additional substances which can create problems in your septic system. Grease, for example, can be a troublemaker. If there is a place for a grease trap in the kitchen, install one; it can help your septic tank do a better job. It is also bad for your septic tank to have rain water or other surface water drain through the system, and, naturally, foreign compounds should be kept out. If you are into photography or other hobbies that require the use of chemicals, never dispose of such chemicals through your septic tank system.

Soil Porosity — Pipe Chart

Minutes for Water To Drop 1 Inch	Absorbtion Field Pipe Need Per Person (in Feet)
1	12
2	15
3	17
4	19
5	20
10	30
15	34
20	40
30	60
60	80

If you have constant problems with your septic tank system, check with your local authority. You may find that you can alter or enlarge the system to eliminate your troubles.

One of the main considerations at the time of the original septic tank installation should have been the water absorption rate of the soil. If the soil porosity is not right, you may never have a good waste-disposal system. You can test the soil by digging several holes with a post hole digger, going down to the level of the disposal field drainage pipes plus about six inches. Fill these holes with water and wait twenty-four hours. Then, add enough water to the holes until you can see about six inches of water. Insert yardsticks in the holes, and time how long it takes for the water level to drop one inch. This interval will help you to determine the size of the disposal field you need to do an adequate job. In addition, the width and depth of the trenches under the pipe — as well as the distance between trenches — can also make a significant difference in the efficiency of your septic system.

The installation of a septic tank can be a do-it-yourself project, but the engineering, planning, and layout must follow the standards established in your community's plumbing code. After you have the municipal authority approve your plans, though, most of what remains to be done is plain old manual labor — digging. If you can dig, therefore, you can save plenty of money. Even if you must rent the digging equipment, you can usually save quite a bit.

A good septic tank system that is properly cared for can be almost problem-free for a long, long time. On the other hand, a bad system or one that is abused can give you trouble every day. If you plan to purchase a home that is equipped with a private sewage-disposal system, it pays to have the old septic tank checked before you buy. The tank could be troublesome and/or inadequate, and you could be buying a system that will give you more headaches than you ever thought possible.

Glossary

Adapter: A fitting used to connect pipes and/or fittings that are of different sizes or of different materials.

Air Chambers: Sections of supply pipes that contain air to cushion the shock waves created when rushing water is suddenly shut off.

Air Gap: The distance between the faucet and the highest level that water can reach in a basin.

Air Vent: See Vent.

Auger: A tool designed to be fed into pipes or fixtures to clean out blockage. Often called a "snake."

Backflow: The potentially dangerous introduction of water from another source into the supply lines.

Branch: Either water or waste offshoots of main pipes that connect to fixtures.

Calking or Caulking: Material, including oakum and lead, used to seal cast-iron pipe joints.

Cap: A solid piece used to close off a pipe end.

Chase: A cutout made in the framing to allow for the installation of pipe.

Cleanout Plug: A threaded cast-iron cap that provides an opening into drains for easier access to clogged areas.

Closet: Toilet; often called a water closet.

Continuous Waste: A waste pipe from more than one fixture, but utilizing only a single trap.

Coupling: A fitting used to connect two pipes.

Developed Length: The total measured length of a run of pipe, including pipe and fittings. Measured along the center line of the pipe.

Diverter: A valve that changes the flow of water from one tap to another (as between a tub and shower).

Dope: Pipe joint compound.

Drain: Any pipe that carries away water and waste.

Ell: Short for elbow. A fitting used for making turns in pipe connections.

Escutcheon Plate: A metal plate used to enclose the point where a pipe or fixture enters the wall or floor.

Female Thread: The end of a fitting with internal threads.

Fittings: Any device used to join sections of pipe together.

Fixture Unit: A unit of measure describing the amount of waste load a particular fixture discharges. A unit is the equivalent of one cubic foot of water per minute.

FPT: Female Pipe Thread.

Frost Line: The depth to which soil freezes.

FSPS: Female Standard Pipe Size.

Gate Valve: A valve that regulates the flow of water through a gate-like disc inside.

Globe Valve: Valve with a spherical globe-like body.

Grade: The slope of a pipe.

Group Vent: Vent that functions for more than one trap.

Hanger: A hanging pipe support.

Horizontal Branch: Any offshoot drain coming off of the vertical stack.

Increaser: A fitting that has a larger opening at one end to allow it to be connected to a larger sized pipe.

Inspection Panel: A plate allowing access to pipes and traps that are inside a wall.

Joint: Any connection between two pipes or parts of a plumbing system.

Joint Compound: A compound put on threaded connections to prevent leaks.

Main: The principal supply pipe to which branches may be connected.

Main House Drain: The pipe that collects waste from all the branches going to fixtures and delivers it to the main house sewer.

Main House Sewer: The pipe that connects your house to the sewer main or to a septic tank.

Main Sewer: See Sewer Main.

Male Threads: The end of a fitting pipe or fixture with external threads.

MPT: Male Pipe Threads.

MSPS: Male Standard Pipe Size.

Offset: A combination of pipe and elbows that move the run of pipe over to make it run parallel to the original pipe.

Pipe Strap: A metal strap used to hang or hold pipe in place on walls, supports, or ceilings.

Pipe Support: Any kind of brace used to hold up pipes.

Pitch: The degree of slope in a run of pipe.

Potable Water: That which is suitable to drink.

Putty: A soft sealer material; not to be confused with glazier's putty used in replacing window panes.

Reducer: A fitting with a smaller opening at one end to allow for connecting to a smaller size pipe.

Risers: The vertical supply pipes that carry water upward into various fixtures in the house.

Rough-In: The complete installation of all plumbing pipes (both supply and drain) along with vents, but not including any fixtures.

Septic Tank: A watertight receptacle that converts sewage to harmless liquid by a bacterial action; used when a community system is not available.

Sewer Main: The community sewage line that takes the waste from private property to the disposal system.

Soil Pipe: Pipes that carry away sewage from toilets.

Stack: All vertical mains that are part of the soil, waste, or vent system.

Tap: A faucet or hydrant for drawing water from a supply line; also, a tool used for cutting internal threads.

Tee: A T-shaped fitting with three openings to receive pipes, fittings, or fixtures.

Trap: A curved portion of a drain pipe designed to retain water that prevents backup of sewer gas without hindering the flow of waste water.

Union: A fitting used for joining pipes that allows for disconnecting without removal of the pipe.

Vent: A pipe that allows a supply of air into the drain system to help control siphoning and back pressure, aiding in the flow of water out of the drain.

Vent Stack: A vertical vent stack.

Water Hammer: The noise that results when air chambers do not control the force of the rushing water as a tap is shut off.

Water Main: The supply pipe that brings water to your house main.

Wet Vent: A pipe that serves for both drainage and venting.